CELL REPRODUCTION

ENCYCLOPAEDIA OF CELL BIOLOGY - III

CELL REPRODUCTION

By

Dr. M.Prakash

Dept. of Zoology
M.M.H. Post Graduate College
Ghaziabad
(U.P.)

DISCOVERY PUBLISHING HOUSE PVT. LTD.
NEW DELHI-110 002

Reprinted – 2019
First Published - 2010

ISBN: 978-81-8356-454-0 (Set)

ISBN: 978-81-8356-561-5

© Author

Cell Reproduction

Published by:
DISCOVERY PUBLISHING HOUSE PVT. LTD.
4383/4B, Ansari Road, Darya Ganj
New Delhi-110 002 (India)
Phone: +91-11-23279245, 23253475; 43596065
E-mail: discoverybooksindia@gmail.com
discoverypublishinghouse@gmail.com
web: www.discoverypublishinggroup.com

Printed at:
Infinity Imaging Systems
Delhi

Preface

The present title "Encyclopaedia of Cell Biology" has been written for undergraduate and postgraduate students of all Indian universities. It is a text on structural and functional unit of the organism that has been thoroughly checked and ornamented with easy and clear illustrations. Special emphasis has been concentrated on the fundamentals of the subject. All related students will find this well established text eminently suitable for introductory courses. Professional cell biologists, researchers, and students may find it useful for updating, and reference purposes. One of our major concerns of this title is to provide students with a basic understanding of what a cell is and why the cell is the fundamental unit of life. Our approach throughout has been to focus on the major question involved and the experimental approaches utilized in addressing these questions. Each important point is illustrated with an example to help clarify the idea; the numerous line drawings have all been chosen or created to help convey particular points. The result is a text that is easy to read, without resorting to verbal gimmicks or talking down to readers.

To make the work more comprehensive and informative, the author has consulted many authoritative books, research journals, abstracts, monographs etc., so there can be no claim to originality except in the manner of treatment.

The author expresses his thanks to his friends and colleagues whose continue inspirations have initiated him to bring out this book.

The author expresses his gratitude to Mr. Wasan and staff of M/s Discovery Publishing House Pvt. Ltd. for their whole hearted cooperation in the publication of this book.

Author

Contents

1

INTRODUCTION

Although some living systems consist of only one cell and others of several billion, even the most complex organism begins life as a single cell, the fertilised egg. In most multicellular organisms, including human beings, this cell divides to form two cells, and each new cell divides again and again, eventually forming the complex tissues, organs, and systems of the developed organism. Like the bricks of a building, cells are the building blocks of the organism. The cell is the smallest unit of living material capable of carrying on all the activities necessary for life.

We might say that it is the smallest structure with a complete metabolism because it has all of the physical and chemical components needed for its own maintenance and growth. When provided with essential nutrients and an appropriate environment, cells can be kept alive in laboratory glassware for many years.

No cell part is capable of such survival. One of the most basic generalisations of biology is the *cell theory*. In 1838 and 1839 two German biologists, Matthias Schleiden and Theodor Schwann, proposed that all living things are made up of cells and cell products, and that the cell is the basic unit of living organisms. The cell theory was extended in 1855 by Rudolf Virchow, who stated that new cells come into existence only by the division of previously existing cells.

Cells cannot arise by spontaneous generation from nonliving matter. In 1880 August Weismann pointed out the corollary of this, that all the cells living today can trace their ancestry back to ancient times.

GENERAL CHARACTERISTICS OF CELLS

A cell consists of jelly-like cytoplasm surrounded by a cell membrane, a physical boundary that separates the cell from the outside environment. Most cells contain a *nucleus* (or *nucleoid* in monerans), which contains the DNA, the chemically coded instructions for synthesizing all the proteins needed for growth, repair, and reproduction. Many other organelles, internal structures with specialised functions, are suspended within the cytoplasm.

Almost all cells are microscopic. An "average" cell measures about 10 micrometers (1/2500 inch) in diameter. This means that if you could line up about 2500 typical cells end to end, the resulting cellular parade would measure less than 3 centimeters (only about 1 inch).

Although they have many features in common, cells vary widely in size and appearance. In fact, the size and shape of a cell are related to its specific function. Epithelial cells, which are specialised to cover body surfaces, look like tiny building blocks. Nerve cells have long extensions that receive or transmit messages long distances through the body. An extension of the sciatic nerve, for example, may extend from spinal cord to foot.

Although such a nerve cell may be more than a meter long, its diameter is so tiny that no part of it can be seen without the aid of a microscope. Certain white blood cells in the body resemble unicellular amebas in their ability to change shape as they flow along from one location to another. Plant cells often have large fluid-filled structures called vacuoles, and they contain chloroplasts.

The largest cells are birds' eggs, which consist largely of yolk that provides nourishment for the developing bird. However, yolk is an inert material that is not really part of the metabolic structure of the cell. Neither the shell nor the white of the bird's egg is considered part of the cell because these structures consist of nonliving material secreted by the mother's reproductive tract.

HOW CELLS ARE STUDIED

Cells are so small that one might well wonder how we know so much about what goes on inside them. The biologist's most important tool for studying the internal structure of cells has been the microscope. Anton van Leeuwenhoek is credited with developing some of the earliest microscopes and with leaving written records of the structures he studied. In 1665 Robert Hooke examined a slice of cork with the aid of a crude, homemade microscope.

Because the tiny compartments he saw reminded him of the little rooms, or cells, of a monastery, he called them cells. What Hooke saw were the cell walls of dead cork cells. In later observations he described cell contents, but it was not until two centuries later that scientists realised that the important part of the cell is its contents and not its outer walls or membranes.

During the 1800s scientists studied various cells and observed a variety of intracellular structures, which they referred to as *organelles* (little organs) because they were thought to perform special jobs within the cell, just as our organs perform specific jobs within our bodies. The ordinary compound light microscope, the kind used by students in most college laboratories, was responsible for the discovery of most cell structures. The *light microscope*, gradually improved since Hooke's time, uses visible light as the source of illumination (see Focus on Viewing the Cell).

During the last three decades the development of the electron microscope has enabled researchers to study the fine detail (ultrastructure) of cells. The *electron microscope* floods the specimen being studied with a beam of electrons rather than with light waves.

Magnification is the ratio of the size of the image to the size of the specimen. Whereas the ordinary light microscope can magnify a structure about 1000 times, the electron microscope can magnify it 250,000 times or more. Another advantage of the electron microscope is its superior resolving power.

Even more important than magnification, *resolving power* is the ability to reveal fine detail; it is expressed as the minimum distance between two points that can be distinguished as separate and distinct points, rather than as one single point. Whereas a light microscope equipped with very fine lenses can resolve objects about 500 times better than the unaided human eye, the electron microscope has a resolving power of more than 10,000 times that of the human eye.

Cell structure has also been studied extensively by physical and chemical methods. Cells can be broken apart and then centrifuged (spun) at high speeds to separate the cellular components. Using a variety of techniques, organelles can be separated from one another on the basis of size or density. Once separated from its surrounding cellular components, a given type of organelle can be analysed biochemically to determine its composition. Another technique used to determine the function of an organelle is to monitor its activities in a test tube under controlled conditions.

PROKARYOTIC AND EUKARYOTIC CELLS

Members of kingdom Monera—the bacteria and cyanobacteria-consist of *prokaryotic cells*, which are fundamentally different from the *eukaryotic cells* of all other organisms. Prokaryotic cells are simpler and generally smaller than eukaryotic cells and are thought to have evolved first. The word prokaryotic means "before the nucleus," and these cells have no nuclear membrane.

Prokaryotic cells do have one or more nuclear regions, sometimes referred to as *nucleoids*, in which DNA is concentrated. Prokaryotic cells also lack other membrane-bound organelles typical of eukaryotic cells. In some prokaryotic cells the cell membrane is folded inward to form a complex of internal membranes along which the reactions of cellular respiration are thought to take place.

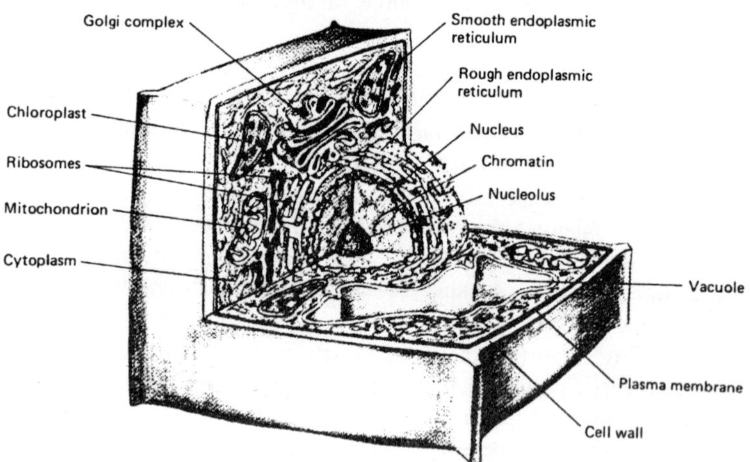

Figure 1.1 : A "typical" plant cell, as visualised by an artist.

Prokaryotic cells that carry out photosynthesis contain the green pigment chlorophyll associated with flat, sheetlike membranes called *lamellae*, but these lamellae are not distinct organelles. As implied by the term *eukaryotic,* which means "good nucleus," or "true nucleus," eukaryotic cells have distinct nuclei surrounded by nuclear membranes.

They also have many types of membrane-bound organelles that partition the cytoplasm into compartments. Table elsewhere in this chapter summarises the types of organelles typical of eukaryotic cells.

Some organelles are found only in specific kinds of cells. For example, chloroplasts, structures that trap sunlight, are found only in cells that carry on photosynthesis. Plant cells differ from animal cells in several ways: (1) although all cells are limited by cell membranes, plant cells are also surrounded by stiff cell walls of cellulose, which limits any change in position or shape; (2) plant cells contain plastids, membrane-bound structures that produce and store food material, the most familiar and abundant of which are chloroplasts; (3) most plant cells have one large or several small conspicuous compartments, called vacuoles, used for transporting and storing nutrients and waste products; and (4) certain organelles-centrioles (and probably lysosomes)-are absent in the cells of complex plants.

INSIDE THE CELL

Early biologists believed that the cell consisted of a homogeneous jelly, which they referred to as protoplasm. With the electron microscope and other modern research tools, perception of the world within the cell has been greatly expanded. We now know that the cell is a highly organised, amazingly complex structure.

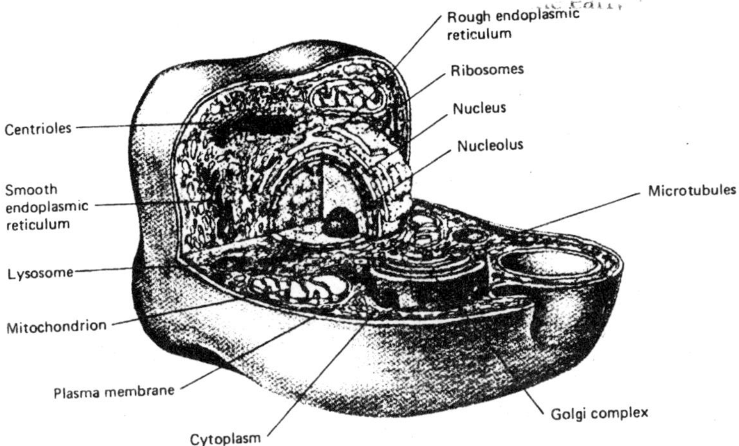

Figure 1.2: An artist's conception of a "typical" animal cell.

It has its own control center, internal transportation system, power plants, factories for making needed materials, packaging plants, and even a "self-destruct" system. Today the word protoplasm, if used at all, is used in a very general way. Specifically, the portion of protoplasm outside the nucleus is called *cytoplasm*, and the corresponding jelly-like material within the nucleus is termed *nucleoplasm*. Within the fluid component of the cytoplasm and nucleoplasm the various organelles are suspended.

Endoplasmic Reticulum and Ribosomes: Internal Transport and Synthesis

A maze of internal membranes extends throughout the cytoplasm of many eukaryotic cells, forming an extensive complex of branching tubules, the *endoplasmic reticulum (ER)*. The ER appears to be continuous with

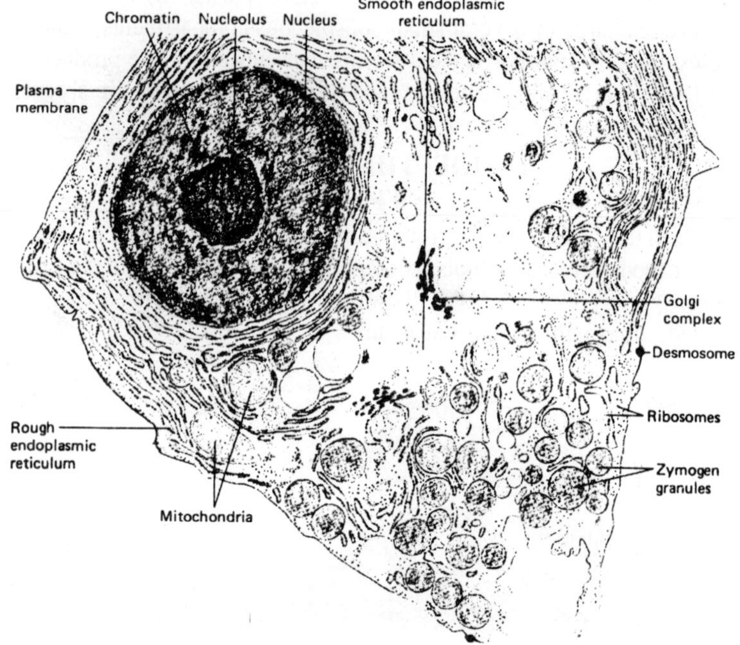

Figure 1.3: The structure of an animal cell.

the cell membrane and nuclear membrane. In electron micrographs, the ER may appear to be discontinuous. This is because such photographs are taken of a thin slice through the cell, but the ER is continuous in three dimensions.

The membranes of the ER divide the cytoplasm into a multitude of compartments in which different types of reactions may occur. In fact, the membranes of the ER contain a variety of enzymes and serve as a framework of surfaces on which many reactions take place. The ER also functions as a system for transporting materials from one part of the cell to another and perhaps to the outside environment as well. Expanded regions of the ER may serve as temporary storage areas for certain substances. Still another important function of the ER is synthesis of some types of compounds.

Two types of ER can be distinguished, smooth and rough. *Smooth ER*, so called because the outer surfaces of its membranes have a smooth appearance, produces steroids in certain types of cells. In other cells it functions in cellular secretion. The smooth ER of liver cells functions in lipid metabolism and also in detoxifying drugs. When an experimental animal is injected with the barbiturate phenobarbital, the

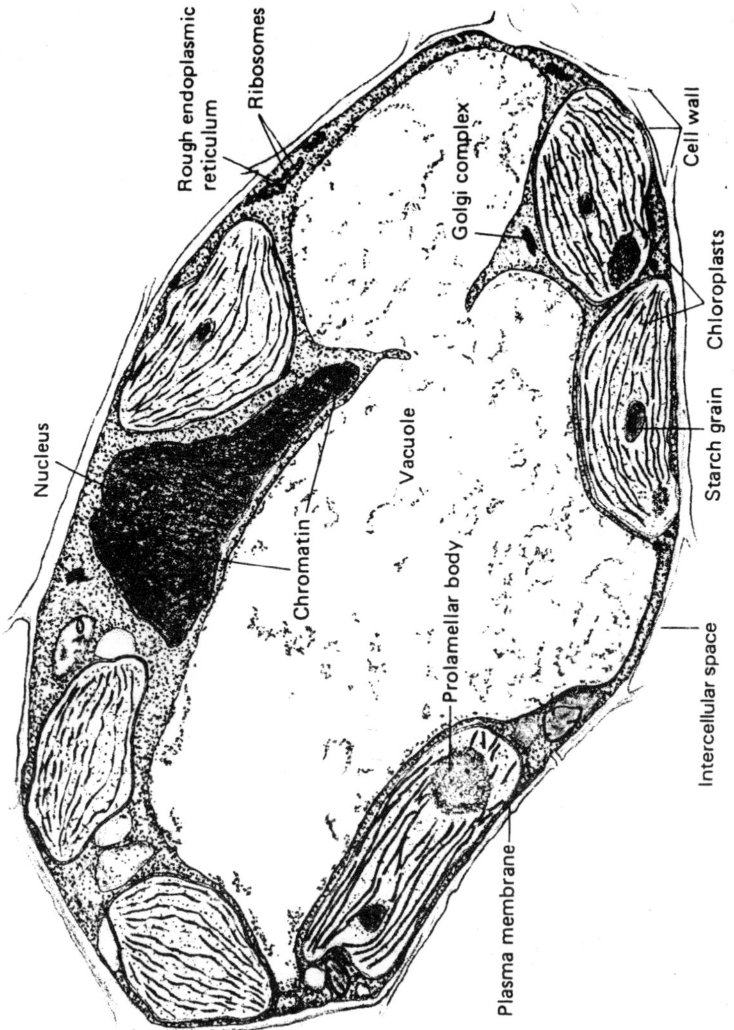

Figure 1.4: The structure of a plant cell.

amount of smooth ER in the liver cells increases over a period of several days. In addition, enzymes known to break down phenobarbital increase in concentration within the smooth ER membranes when this drug is administered over a period of time.

Rough ER has a granular appearance due to the presence of minute organelles called ribosomes that stud its outer walls. *Ribosomes* are the site of protein synthesis. Rough ER is especially extensive in cells that

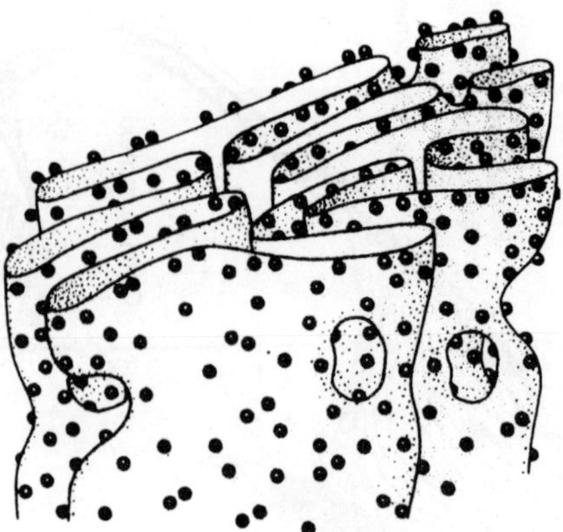

Figure 1.5: Rough endoplasmic reticulum.

synthesize proteins for export from the cell. Not all ribosomes are attached to the ER. Some float freely in the cytoplasm. These assemble proteins for use inside the cell.

Ribosomes are found in all kinds of cells, from bacteria to complex plant and animal cells. Composed of RNA and protein, a ribosome consists of two subunits that are combined to form the complete protein-synthesizing ribosome. In many cells, clusters of five or six ribosomes termed polysomes appear to function in protein synthesis.

The Golgi Complex: A Packaging Plant

The *Golgi complex*, first described in 1898 by the Italian investigator Camillo Golgi, consists of stacks of platelike membranes. The membranes may be distended in certain regions forming vesicles filled with cell products. In animal cells the Golgi complex is usually located at one side of the nucleus.

The Golgi complex functions mainly as a packaging plant and is most highly developed in cells that are specialised to secrete products. Proteins manufactured along the rough ER pass to the Golgi. They are released from the ER sealed off in little packets of membrane, or *vesicles*. These vesicles fuse with the older membranes of the Golgi complex.

Within the Golgi complex the protein secretion may be concentrated by the actions of its membranes. During storage proteins may be modified. Often some carbohydrate component is added to form a glycoprotein. (Polysaccharides may be synthesized from simple sugars within the Golgi complex.) A vesicle containing the modified product is released from the Golgi complex. In cells that secrete substances,

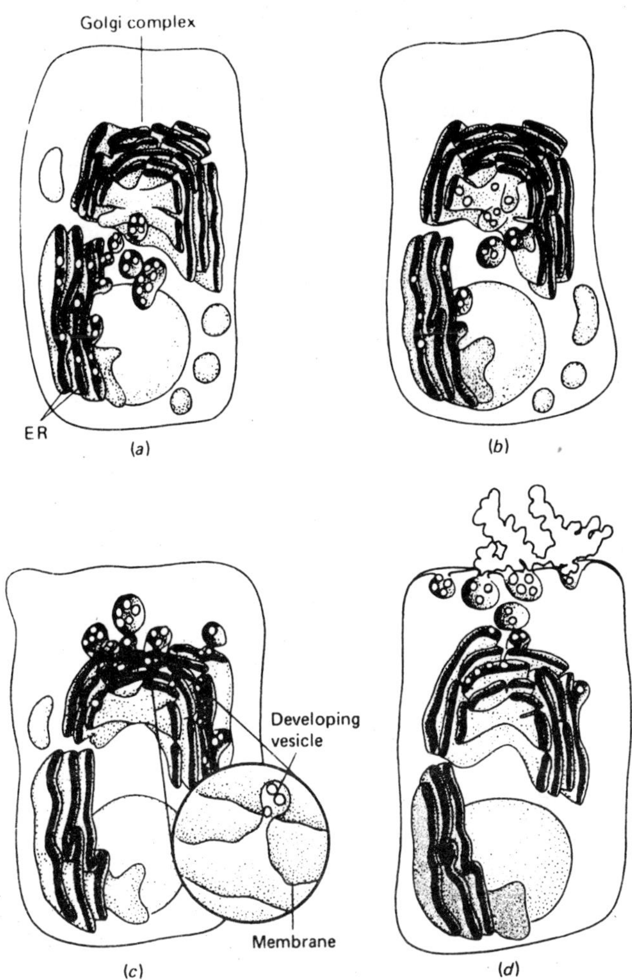

Figure 1.6: The Golgi complex, Diagram.

such as gland cells, the vesicles fuse with the cell membrane and release their contents to the exterior of the cell. Although it is especially prominent in cells specialised to secrete products, the Golgi complex also performs an important function in nonsecreting cells: It packages intracellular digestive enzymes in little organelles called lysosomes.

Lysosomes: Breaking Down Harmful Materials

Small sacs of digestive enzymes called lysosomes are released from the Golgi complex and dispersed throughcut the cytoplasm. With their potent enzymes, lysosomes break down worn-out organelles and foreign matter into their component molecules. In a cell that is short of fuel, lysosomes may break down organelles so that their component molecules can be used as fuel. When a white blood cell ingests a bacterium or some other scavenger cell ingests debris or dead cells, the foreign matter is enclosed in a vesicle composed of part of the cell membrane.

One or more of the cell's lysosomes then fuse with the vesicle containing the foreign matter. The lysosome pours its powerful digestive enzymes into the vesicle, destroying the material within it. These enzymes can digest almost any type of large molecule found in cells.

The lysosomal membrane itself is able to resist the digestive action of these powerful enzymes, but this is thought to require a continuous expenditure of energy. When a cell dies, the lysosome membrane breaks down, releasing digestive enzymes into the cytoplasm, where they break down the cell itself. This "*self-destruct*" system accounts for the rapid deterioration of many cells following death.

Some forms of tissue damage as well as the aging process may be related to leaky lysosomes. Rheumatoid arthritis is thought to result in part from damage done to cartilage cells in the joints by enzymes that have been released from lysosomes. Cortisone-type drugs, which are used as anti-inflammatory agents, stabilize lysosome membranes so that leakage of damaging enzymes is reduced.

Microbodies

Microbodies are membrane-bound organelles that contain a variety of oxidative enzymes rather than digestive enzymes. These structures contain enzymes that promote an assortment of metabolic reactions, such as the conversion of fats to carbohydrates. During some of these reactions, hydrogen peroxide, a substance toxic to the cell, is produced. *Peroxisomes*, the type of microbody in which these reactions occur, also contain enzymes that split hydrogen peroxide, rendering it harmless.

MITOCHONDRIA: POWER PLANTS OF THE CELL

Often referred to as the power plants of the cell, the *mitochondria* (singular, mitochondrion) are the sites of cellular respiration. More than 1000 mitochondria have been counted in a single liver cell, but the number varies among cell types. In general they are more abundant in cells that are very active metabolically. Mitochondria are capable of changing size and shape, of fusing with other mitochondria to form larger structures, or of splitting to form smaller ones. They may appear

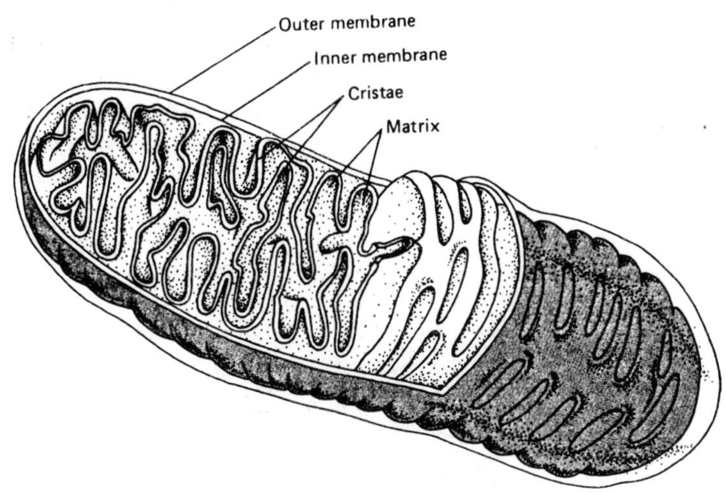

Figure 1.7: The mitochondrion.

as spheres, rods, sausages, or threads. The mitochondrion is bounded by a double membrane. Both the outer and inner membranes consist of lipid bilayers in which a variety of protein molecules are embedded.

The outer layer of the mitochondrial membrane is smooth, but folds of the inner layer, called *cristae*, project into the interior of the mitochondrion. Cristae serve to increase the available surface area, and some of the enzymes needed for cellular respiration are organised along these folds. Other enzymes are found in the semifluid material within the inner compartment (matrix). Each mitochondrion has a small amount of DNA, enough to code for about 15 proteins, and also contains ribosomes. Some proteins are indeed synthesized within these organelles.

Plastids: Energy Traps and Storage Sacs

Structures known as plastids produce and store food materials in algae and plant cells. *Chloroplasts*, the most common type, contain the green pigment *chlorophyll*, which traps light energy for photosynthesis. Chloroplasts also contain a variety of yellow and orange pigments known as carotenoids. Although a unicellular alga may have only a single large chloroplast, cells of complex plants often possess 20 to 40 of these interesting organelles.

The chioroplasts of plant cells are typically disc-shaped structures bounded by an inner and outer membrane. Inside the chloroplast there are parallel platelike membranes called lamellae. At various points along the lamellae the membranes are expanded to form disclike structures called *thylakoids*. Chlorophyll is present within the thylakoids, and the light-dependent reactions of photosynthesis (those involved in capturing the energy of sunlight) take place within these structures. The thylakoids are arranged in stacks called *grana* (singular, granum). Like mitochondria, chloroplasts contain DNA and ribosomes and manufacture some proteins. They are also able to grow and divide to form daughter chioroplasts. Details of chioroplast structure and function will be discussed in other chapter of this book.

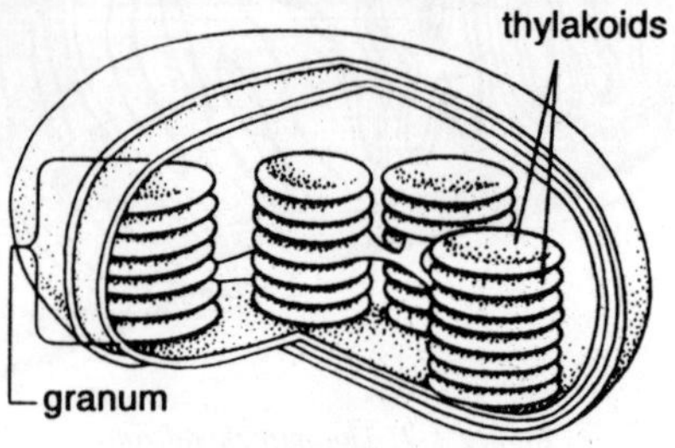

Figure 1.8: The chloroplast.

Plant cells also contain colourless plastids termed *leukoplasts* which serve as centers for the storage of starch and other materials. Starch is commonly deposited in leukoplasts present in storage roots and stems and in seeds. A third type of plastid, the *chromoplasts*, contains pigments that give colour to flowers, autumn leaves, and ripe fruit.

MICROTUBULES AND MICROFILAMENTS: CELL SHAPE AND MOVEMENT

Microtubules are very small, hollow cylinders composed mainly of proteins called tubulins. Microtubules can grow by the addition of more tubulin subunits or can shorten by the disassembly of subunits. Microtubules located just beneath the cell membrane determine the shape of many cells.

These structures also play a role in some types of cell movements. For example, they help move chromosomes during nuclear division. In nerve cells microtubules are important in the rapid transport of substances down the axon (the elongated extension of a nerve cell).

Microtubules are the main structural components of cilia and flagella, organelles that are discussed in a later section. Solid cytoplasmic *microfilaments* play additional roles in cell structure and movement. They are composed of the proteins actin (or actin-like substances) and myosin, both important in muscle contraction.

Microfilaments are essential in cellular movement, such as the flowing of cytoplasm that enables a cell to move from one place to another. They are also involved in cytoplasmic streaming which results

Figure 1.9: Microtubules.

in the continuous motion of organelles characteristic of plant cells.

The Microtrabecular Lattice: Intracellular Framework

The fluid portion of the cytoplasm (the cytosol), long thought to be homogeneous and structureless, actually contains an irregular threedimensional network of extremely slender protein threads. Known as the *microtrabecular lattice*, this network extends throughout the cytoplasm and is attached to the cell membrane.

Microtubules, microfilaments, endoplasmic reticulum, mitochondria, and other organelles are integrated with the lattice and suspended in it. In fact, the lattice links the microtubules, microfilaments, and organelles into a structural and functional unit.

The microtrabecular lattice changes in shape and contracts in various regions, continuously redistributing the organelles, and is perhaps responsible for many cellular responses. For example, the movement of pigment granules within the pigment cells in the skin of fishes, amphibians, and reptiles permits the animal to lighten or darken its skin colour rapidly to match the colour of its surroundings. Such movement, which is under hormonal and nervous control, is accomplished rapidly by the microtrabecular lattice. Shortening and thickening of microtrabecular threads results in movement of the pigment granules to the center of the cell. Elongation of the fibers disperses the pigment granules throughout the cell.

Centrioles

Many types of cells possess a pair of tiny *centrioles*, organelles that function in nuclear division. A centriole is a hollow cylinder composed of nine triple microtubules. Centrioles are located within a dense area of cytoplasm, the centrosome, which is generally located near the nucleus.

Cilia and Flagella

Many cells have movable whiplike structures projecting from the free surface. If a cell has one, or only a few, of these appendages and they are relatively long in proportion to the size of the cell, they are termed *flagella*. If a cell has many short appendages, they are called *cilia*.

Flagella and cilia are very similar in structure, and both function either in moving the cell along through a watery environment or in moving liquids and particles across the surface of the cell. Flagella or cilia are commonly found on one-celled and small multicellular organisms and on the sperm cells of animals.

They are the principal means of locomotion of such cells. Cilia

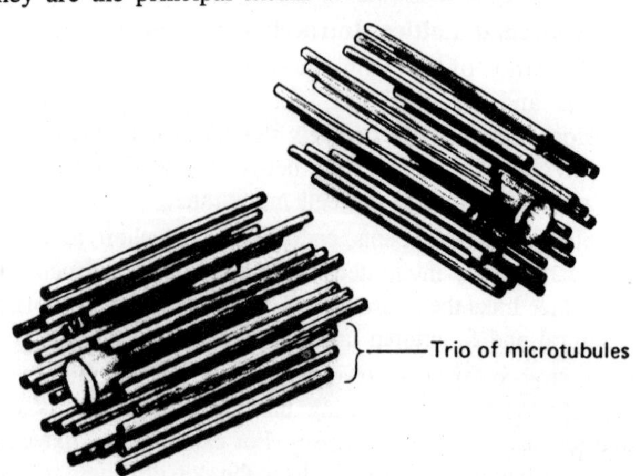

Figure 1.10: Centrioles.

commonly occur on the cells lining the internal ducts of animals; their beating assists in moving materials through these passageways (e.g., through the respiratory passageways). Each cilium or flagellum consists of a slender, cylindrical stalk covered by an extension of the cell membrane and containing a cytoplasmic matrix with groups of microtubules embedded in it.

Nine pairs of microtubules are arranged around the circumference, -and two single microtubules are located in the center. This 9 + 2

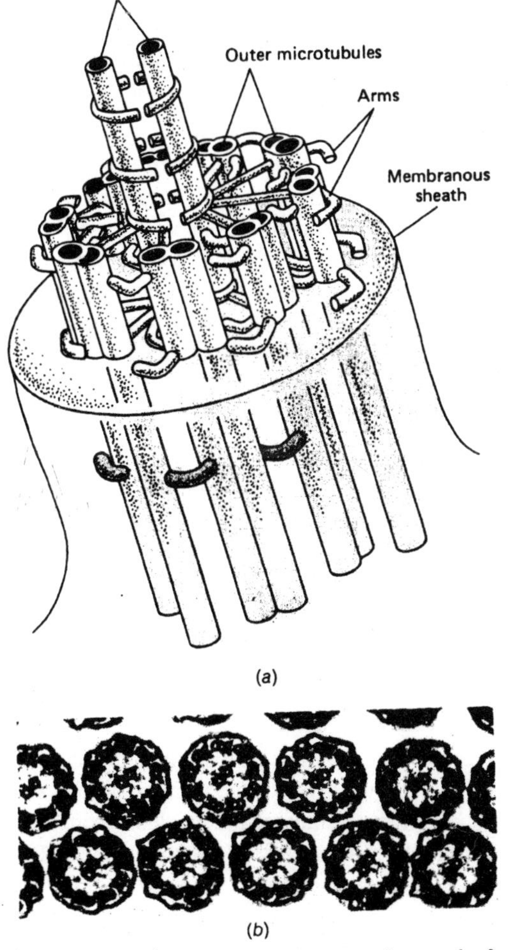

Figure 1.11: Cilia (a) structure of a cilium. (b) Electron micrograph of cross sections through cilia showing the 9 + 2 arrangement of microtubules.

arrangement is characteristic of all cilia and flagella except those found in organisms that belong to kingdom Monera.

Microtubules are thought to be important in the movement of the cilium or flagellum, perhaps by complex sliding movements along one another that bend the entire organelle. At the base of the stalk of a cilium or flagellum there is a *basal* body, a structure similar to a centriole. The basal body is the structure from which the stalk arises. Cilia develop as the cell matures.

VESICLES AND VACUOLES

We have already described several types of vesicles, for example, those that are released from the Golgi complex filled with a secretory product. The term vesicle is used to describe many kinds of small membranous bags found within the cell.

Although the terms vesicle and vacuole are sometimes used interchangeably, vacuoles are actually larger structures, sometimes produced by the merging of many vesicles. A vacuole may be defined as a bubblelike space usually filled with watery fluid and bordered by a membrane. Present in many types of cells, vacuoles are most common in plant cells and the cells of protists. Most protozoa have food vacuoles containing food undergoing digestion, and many have contractile vacuoles, which remove excess water from the cell.

More than half the volume of a plant cell may be occupied by a large central vacuole containing stored food, salts, pigments, and wastes. Plants lack waste disposal systems and often utilize vacuoles as a storage place for toxic materials. Such waste products often aggregate and form small crystals.

THE CELL NUCLEUS

Of all the structures within the cell, the nucleus is the most prominent. Perhaps for this reason investigators guessed that the nucleus served as the control center for the cell even before techniques to prove that hypothesis were available.

During recent years all kinds of experiments have been performed that have confirmed the vital role of the nucleus. In one such experiment the researcher surgically removed the nucleus from a living ameba. The enucleated ameba was unable to eat or grow, and it died after a few weeks.

However, if after a day or two an enucleated ameba was given a new nucleus, it made a complete recovery. These and other experiments show that the nucleus is essential to the well-being of the cell.

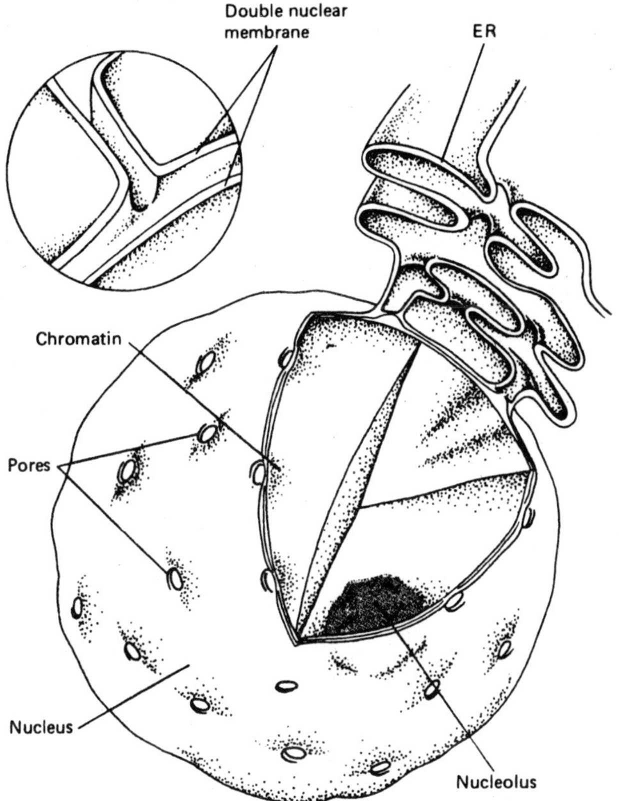

Figure 1.12: The cell nucleus. A portion of the nucleus has been cut away to show the interior.

The Nuclear Envelope

The nucleus is separated from the surrounding cytoplasm by a *nuclear envelope*, a double membrane that regulates the flow of materials into and out of the nucleus. In some cells the outer nuclear membrane appears to be continuous with the endoplasmic reticulum and the Golgi complex.

The two membranes of the nuclear envelope are fused at intervals forming *nuclear* pores-channels through which the interior of the nucleus communicates with the cytoplasm. However, passage of materials through the pores is thought to be selective.

The Nucleolus

The nucleus may contain one or more large, prominent *nucleoli* (*singular, nucleolus*). The nucleolus is not walled off from the rest of the nucleus by any kind of membrane. Rather, it is a dense region rich in RNA and protein-materials used to manufacture ribosomes. A nucleolus develops around a certain region of one chromosome, the nucleolar organizer, which contains the genes for synthesizing ribosomal RNA (the type of RNA found in ribosomes). During nuclear division, the nucleolus becomes disorganised and its components disperse. When division is completed, the nucleolus reorganizes.

Chromatin and Chromosomes

In a cell that is not dividing, an irregular network of granules and strands termed *chromatin* is evident. The chromatin consists of protein and uncoiled DNA that may be actively synthesizing substances needed by the cell. When a cell begins the process of nuclear division (mitosis), the chromatin coils and condenses into discrete rodshaped bodies, the *chromosomes*. (You might think of the condensing of chromatin to form a discrete chromosome as being somewhat like taking two intertwined Slinky toys that have been stretched out as far as they can extend, then releasing them slowly to recoil into a small, tightly wound double helix.)

Each chromosome contains several hundred *genes* arranged in a specific linear order; the genes, in turn, are composed of the nucleic acid DNA. Chemically coded within the DNA of the genes are instructions for producing all the proteins needed by the cell.

These proteins determine what the cell will look like and what functions it will perform. The chromosomes serve as a chemical cookbook for the cell, whereas the genes might be compared to the individual recipes. When condensed, the chromosomes may be compared to a closed cookbook:

The recipes are all inside but the pages cannot be read. When the chromosomes elongate forming chromatin, the book is open and the instructions can be followed.

2

SOMATIC CELL REPRODUCTION

The ability to grow and reproduce is a fundamental attribute of cells-one on which the survival of every species of unicellular and multicellular organism is absolutely dependent. The study of cell reproduction is important in many areas of cell biology, including the development and functioning of every organism. A human, for example, begins existence as a single cell, the fertilised egg, and grows by cell reproduction to more than one hundred trillion cells (10^{14}) in adulthood.

To maintain the functions of various tissues, about one trillion (10^{12}) cell divisions occur in an adult human every 24 hours, or 10^7 cell divisions per second. The precursor cells in the bone marrow that differentiate into red blood cells produce 2.5×10^6 new cells per second. Blood contains fewer white blood cells, but their life spans are measured in days instead of months, and they must be replaced by a similarly high rate of cell division.

Epithelial cells that line the intestinal tract and cells in the generative layer of the skin also reproduce at high rates to replace dead cells. In some other tissues, such as those of the liver, kidneys, lungs, and bones, the rate of cell reproduction is much lower. Two major kinds of cells-skeletal muscle cells and nerve cells-do not reproduce in adults and cannot be replaced. Reproduction of a cell is a momentous accomplishment.

It starts with growth, which is the coordinated increase in the amounts of the thousands of different kinds of molecules that make up all of the parts and functions of a cell, and culminates with partitioning of cytoplasm and chromosomes into two daughter cells by cell division. Despite its complexity, cell reproduction is generally a mistake-free process, having been highly polished by several billion (10^9) years of evolution.

An essential prerequisite for cell reproduction is chromosome replication. Chromosome replication must be coordinated with cell growth, and both must be coordinated with cell division. How this is achieved is one of the major areas of study in cell biology.

Cell reproduction is different enough in prokaryotes and eukaryotic cells to warrant separate discussions. The following section summarises main features of the reproductive process in prokaryotes.

REPRODUCTION OF PROKARYOTIC CELLS

A bacterial cell reproduces by increasing its size, simultaneously replicating its chromosome, and then dividing into two cells, each of which receives one of the daughter chromosomes. Growth from division to division consists of a doubling in the amounts of all components-ribosomes, plasma membrane, cell wall, enzymes, water, and other molecules. A new daughter cell of *E. coli*, for example, begins its independent existence immediately after cell division; it is then a cylinder one μm in diameter and two μm in length.

The diameter remains constant, and the cell doubles its size by growing to four μm in length. During this growth the number of ribosomes increases from 20,000 to 40,000 with a concomitant doubling in the cell's capacity for protein synthesis. In a rapidly growing bacterial cell a major portion of the cell's ATP is used for production of ribosomes.

Cell growth, chromosome replication, and cell division are coordinated as the cell moves through its cycle, but the molecular basis for the coordination is still poorly understood. Measurements of cell size suggest that initiation of DNA replication is triggered when a cell reaches a threshold size, which is well below that reached just prior to division. The DNA replicates as the cell continues to grow. Division occurs at a fixed time interval after completion of DNA replication, by which time the cell has also completed its growth to a doubled size.

A connection between cell growth and initiation of DNA replication has been demonstrated in a variety of experiments. If protein synthesis is blocked with an inhibitory drug like puromycin or by excluding from the nutrient medium an amino acid that the cell cannot synthesize, initiation of DNA replication is prevented. Blocking growth in other ways, for example, by depriving a cell of sources of nitrogen, phosphate ions, or energy (glucose and organic molecules), produces the same result. Thus, if ATP production in *E. coli* falls because of a lack of energy-yielding carbon molecules in the medium, ribosome production stops, cell growth ceases, and DNA replication is not initiated.

In prokaryotes, cell division consists of a separation of the two daughter chromosomes into two distinct nucleoids, then the migration of the nucleoids toward the two ends of the cell, followed by an inward growth of the plasma membrane and cell wall to form a septum across the cell between the nucleoids. Subsequently, a split forms in the plane of the septum wall, and the two daughter cells separate.

The rate of growth of the individual cell, and therefore the rate of cell reproduction, depends on nutrient conditions and temperature. E. *coli* can grow in a medium consisting of a simple organic molecule such as glucose as a source of energy and carbon and various inorganic ions. In nutrient medium with glucose as the carbon and energy source, a cell doubles in size and divides at 37°C every 40 minutes on the average.

In medium rich in amino acids, purines, pyrimidines, and other organic molecules cells grow faster, and the division time (the generation time) is reduced to 20 minutes. This is a remarkably high rate of cell proliferation. A single bacterium reproducing with a generation time of 20 minutes produces a population of 2^{24} or over ten million cells in 8 hours. As nutrients in a culture are used up, individual cell growth slows and then stops.

Cell reproduction ceases, the number of cells remains stationary for a time, and eventually cells begin to die if not transferred into fresh medium. It is often convenient to record the proliferative behaviour of cells in culture with a growth curve, as shown in Figure elsewhere in this chpater. Growth of a culture is initiated by inoculating cells from a dense culture into fresh medium (time 0).

Proliferation begins after a lag of some minutes, during which cells adjust to the fresh medium. Following the *lag phase* the cells multiply exponentially. For this reason the log of cell number plotted against time gives a straight line. The generation time can be read from the slope of the line as the time taken for the population to double.

This period of rapid growth, called the *log phase* or *exponential phase*, is followed by decelerating growth and then a *stationary phase*, in which the cell number does not change. Finally, the culture enters the *decline phase* as the cells begin to die.

REPRODUCTION OF EUKARYOTIC CELLS

The simplest description of cell reproduction is a cycle of growth and division. The cell grows during the period between cell divisions, the interphase, and then divides; the daughter cells repeat the cycle, and the number of cells increases exponentially.

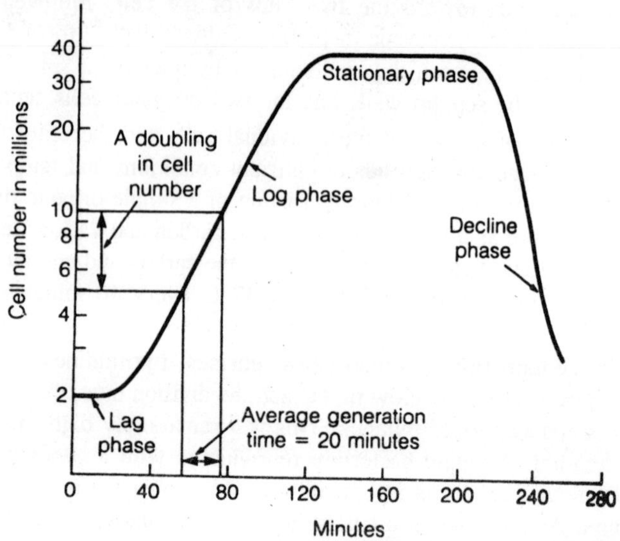

Figure 2.1: Growth curve for a bacterial cell population in rich nutrient medium.

This description of the cell cycle is expanded by marking out the period of DNA replication within interphase. Because the timing of DNA replication in the cell cycle is crucial to the understanding of cell reproduction and its regulation, methods for determining timing are described next.

DNA Replication in the Cell Cycle

The most sensitive method for determining when DNA replication takes place is by autoradiographic detection of incorporation of ^{3}H-thymidine. Since thymidine is used exclusively for DNA synthesis, only those cells engaged in replication incorporate ^{3}H-thymidine and become radioactive. At any instant the cells in a growing population will be in all different parts of the cell cycle, that is, the cell cycles will be asynchronous. Most cells will be in one or another part of interphase; others will be in various stages of division. When ^{3}Hthymidine is given to the cells for a few minutes and the cells are then processed for autoradiography, those that are replicating their DNA (in the S period) will produce an autoradiographic image. This autoradiograph shows that (1) cells in division do not incorporate ^{3}H-thymidine and hence cells in division are not replicating DNA and (2) only some of the interphase cells are replicating DNA, and therefore DNA replication, or the S period, occupies only part of interphase. Where in interphase replication occurs is shown by the following experiment.

An asynchronously reproducing culture of animal cells is given ^{3}H-thymidine for a few minutes to label those cells that are in the S period at the time. Instead of processing the cells for autoradiography as in Figure elsewhere in this chapter, the cells are given fresh medium containing nonradioactive thymidine and no ^{3}H-thymidine so that no further labeling of DNA takes place.

The cells are allowed to continue cycling, but from now on any cell that enters the S period will not become labeled. Further, cells that incorporated ^{3}H-thymidine during the short exposure and then leave the S period, having completed replication, will still be labeled in their DNA.

One hour after removal of ^{3}H-thymidine from the medium a sample of the cells is examined by autoradiography. Particular attention is given to cells in mitosis, and none of these are found to be radioactive-only interphase cells show label. This means that one hour is not long enough for any of the cells that were in the S period at the time of ^{3}H-thymidine labeling to reach mitosis, that is, DNA replication must end at least one hour before mitosis.

A sample of cells examined two hours after removal of ^{3}H-thymidine from the medium gives the same result, which means that DNA replication must end at least two hours before mitosis. In a sample of cells examined at three hours after removal of ^{3}H-thymidine from the medium, 50 percent of the cells in mitosis are found by autoradiography to be radioactive.

These cells must have been in the last minutes of the S period when the ^{3}H-thymidine was available. In the subsequent three hours they finished DNA replication and progressed into mitosis. Half of the mitotic cells in the three-hour sample are not radioactive and must have already finished replication when the

^{3}H-thymidine was available. This means that cells progress from the end of the S period to mitosis at slightly variable rates: some take a little longer than three hours, some take a little less time. The average is calculated as the time required for the fraction of mitotic cells that is labeled to reach 50 percent, in this case, three hours.

In a sample of cells taken at four hours after the short exposure to ^{3}H-thymidine almost 100 percent of cells in mitosis are labeled, which means that almost all cells were replicating their DNA four hours before mitosis. Plotting the percentage of labeled mitotic cells against time after exposure to ^{3}H-thymidine on a graph shows the relationship of DNA replication to mitosis. In samples of cells taken at longer times after exposure to ^{3}H-thymidine all the mitotic cells are labeled, showing that DNA replication takes place continuously over many hours of interphase. Eventually, in cells taken at 11 hours after ^{3}H-thymidine exposure, some of the mitotic cells are not labeled.

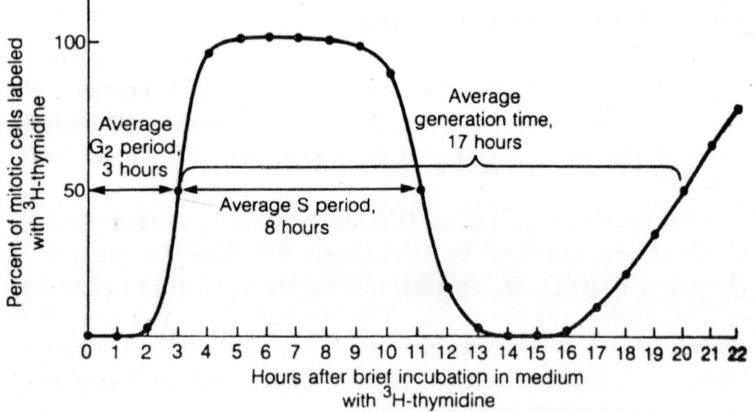

Figure 2.2: The labeled mitosis method for determining the length of the G2 period, the S period, and the generation time.

These are cells that had not yet entered DNA replication when ^{3}H-thymidine was available, that is, cells in the prereplicative part of interphase. The point on the curve in Figure elsewhere in this chapter at which the percentage of mitotic cells that is labeled falls to 50 percent marks the average time preceding mitosis when DNA replication started, or at 11 hours in Figure elsewhere in this chapter.

In cell samples taken at longer times none of the mitotic cells are labeled because none had reached DNA replication at the time ^{3}H-thymidine was available. Eventually, in cell samples examined at longer and longer times, labeled mitotic cells again appear. These are cells that were in DNA replication at the time of availability to ^{3}H-thymidine and have had time to reach mitosis, divide, transit the entire next interphase, and reach the next mitosis.

The information in the graph in Figure elsewhere in this chapter can be used to construct a diagram showing the arrangement of events in the cell cycle. Replication of DNA occurs over an eight-hour interval, the S period, that ends about three hours before mitosis. The threehour interval between the end. of the S period and the start of mitosis is called the *G2 period* (G for gap of time). The interval between mitosis and the next initiation of DNA replication is called the *G1 period.* Cell division takes about one hour in animal cells (determined by microscopic observation) and is designated as M (mitosis) or D (division).

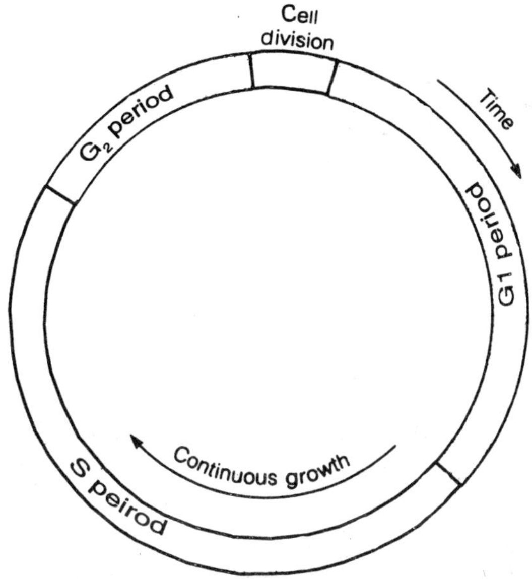

Figure 2.3: The cell cycle and its four major sections, the G1, S, G2 and division periods.

In Figure elsewhere in this chapter, 50 percent of mitotic cells are labeled three hours after exposure of the population to ^{3}H-thymidine. If allowed to continue, the labeled mitotic cells would be detected as labeled cells in mitosis at the end of the next cell cycle. They are responsible for the second rise in labeled mitotic cells in Figure elsewhere in this chapter, reaching 50 percent labeling at 20 hours. The cells reached the first mitosis on average three hours after labeling and reached the second mitosis on average at 20 hours after labeling. Therefore, the time from one mitosis to the next, or the average generation time, is 20 - 3 = 17 hours.

The data in Figure elsewhere in this chapter reveal the generation time (GT), the length of the G2 period, and the length of the S period. The time taken for mitosis (one hour) is known from microscopic study.

The length of G 1 is then calculated: G1 = GT - G2 - S - M, in this case five hours, as shown in Figure elsewhere in this chapter.

This analysis shows that the cell cycle consists of four successive periods that are defined by the occurrence of DNA replication and mitosis. The analysis of these four periods and their interrelations forms the basis of much current research on how cell reproduction is achieved and how it is regulated.

Flow Microfluorimetry (FMF)

Another way to analyze the cell cycle is to measure the amount of DNA in individual cells in an asynchronously growing population. In a completely asynchronous population, at any instant cells are distributed randomly in all parts of the cell cycle. Some cells in the population have not started replication and contain the 2C (diploid) value of DNA.

The C value is the quantity of DNA in a cell with a haploid set of unreplicated chromosomes. Therefore, 2C is the amount of DNA in a diploid cell before DNA replication starts, that is, a diploid cell in GI. Some cells in the growing population contain a 4C amount of DNA, which is the amount expected in a cell in the G2 period or in mitosis before completion of cell division.-The rest of the cells have DNA contents between the 2C and 4C amounts; these are in the process of replicating DNA.

The quantity of DNA in a large number of individual cells can be quickly determined by staining the cells with a dye that binds specifically to DNA. The amount of dye bound in a cell is then proportional to the amount of DNA in the cell.

The kinds of dyes used for this technique are fluorescent: they emit light of one colour when excited by light of another colour. The intensity of emitted fluorescent light is proportional to the amount of dye that is present. Therefore, when a cell whose DNA has been stained with a fluorescent dye is exposed to a light beam of the right colour, it fluoresces with a light intensity that is proportional to its DNA content.

In the technique called *flow microfluorimetry*, cells are fixed with alcohol or another fixative and then stained with a fluorescent dye. The cells are suspended in a salt solution and forced to pass rapidly in single file through a fine capillary tube positioned in the path of a laser beam. As each cell passes through the capillary, the fluorescent dye bound to its DNA is excited to emit light in all directions. The intensity of emitted light is measured by a photoelectric cell positioned out of the path of the laser beam. Cells can be forced through the capillary at a rate of several thousand per second. The light emitted by each one is measured separately and recorded electronically by the flow microfluorimeter (FMF).

Another component of the flow microfluorimeter compiles the individual cell measurements in a display called a histogram, in which the number of cells with a given amount of DNA is plotted against the DNA amount. The histogram in Figure elsewhere in this chapter shows the DNA contents of tens of thousands of cells for an asynchronously growing population. A peak in cell number is present at the 2C amount of DNA, representing cells in GI. Another peak, at the 4C amount of DNA, represents cells in G2 and mitosis. Cells with intermediate amounts of DNA are S-period cells.

If the cells in a growing population are distributed randomly throughout the cell cycle, which is the case for a population that is growing completely asynchronously, then the percentage of cells in any given segment of the cell cycle is proportional to the duration of that segment. Thus, the percentage of cells in G1 is proportional to the length of the G 1 period, the percentage of cells in the S period is proportional to the length of the S period, etc.

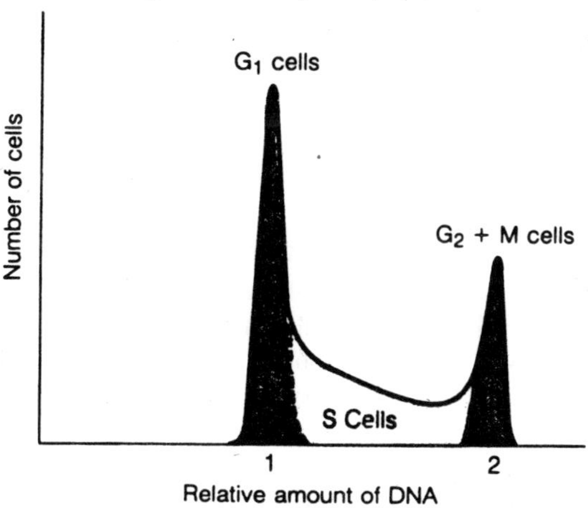

Figure 2.4: The DNA content per cell for a population of cells determined by flow microfluorimetry. The dashed lines represent resolution of the cell population into three components, Gl cells, S period cells, and cells in G2 and M.

In FMF analysis the percentages of cells in the G1, S, and G2 + M periods can be calculated by resolving the histogram into the three corresponding component curves. The areas under the three component curves give the relative percentages of G1, S, and G2 + M cells. However, because there are always about twice as many cells entering G1 as there are cells leaving G2 in a logarithmically expanding population, with intermediate numbers of cells between the two extremes, a correction factor is applied in order to obtain accurate measures of the sizes of the G1, S, and G2 + M populations:

When the percentages of Gi, S, and G2 + M cells have been determined, the relative durations of G1, S, and G2 + M are also known. If the average generation time for cells in the population is also known, for example, from a population growth curve, then the absolute length of each period can be determined.

Since M usually lasts one hour, the length of G2 is simple to calculate. For example, if 40 percent of the cells are in G1, 40 percent in S, and 10 percent in G2 + M, and the generation time is 20 hours, then, as an approximation, G 1= 8 hours, S = 8 hours, G2 = 3 hours, and M = 1 hour. To make these estimates a bit more accura corrections can be introduced to account for the disc bution of ages of cells in the cycle, as mentioned above.

Analysis of the cell cycle by FMF is not quite accurate as analysis by ^{3}H-thymidine and autoradiogi phy because incorporation of small amounts of ^{3}H-th midine can be detected more readily than small increases in DNA content. However, the FMF method has

the advantages of speed and convenience in analysis cultured cells. For analysis of cell tissues in a multicf lular organism the autoradiographic method ordinari has to be used since the FMF requires that cells I suspended in a solution.

Lengths of the Gi, S, G2, and M Periods

For mammalian cells, both in culture and in a organism, the lengths of the S, G2, and M periods ten to be constant in all cells. The S period lasts about hours, G2 about 3 hours, and M about one hour. The length of G1 is far more variable from cell to cell. Fe cultured mammalian cells it usually ranges from 3 to 1 hours, depending on the species and on the type of cell.

Within an organism, the length of G1 differs mud more from one cell type to another. In adult tissues G is usually much longer than in cultured cells. For example, in the epithelial cells that line the digestive tract, which are among the most rapidly reproducing cells in the body, the length of G 1 ranges from a few hours is the small intestine to more than 100 hours in the esophagus. The most rapidly reproducing cells in ani mals generally are those of early cleavage-stage embryos;

in these the G1 period is totally absent, and suc. cells proceed from mitosis directly into the next S period.

In short, the differences in the length of G 1 give ris to the main differences in cell cycle times and therefor determine differences in rates of cell reproductior More about the length of G1 is included later in th chapter in connection with regulation of cell reproduction in normal cells and in connection with cancer cells.

The cell cycles for cells in other animals and in plants are similar to those in mammals. In some rapid reproducing, unicellular organisms the generation times are much shorter (e.g., 1.5 hours in yeast), but the arrangement of G 1, S, G2, and M is much the same. In others, such as *Amoeba proteus* and *Physarum,* the cycles are different. The G1 period is totally absent and G2 is the longest part of the cycle.

Progression through the Cell Cycle

The cell progresses through a sequence of stages to accomplish its reproduction. This section briefly describes the progression.

The GI period

The completion of cell division marks the end of one cell cycle and the start of the next for the two daughter cells. With the few exceptions already noted above, the cycle begins with entry into the G1 period. During G1 a cell grows continuously, as it does during S and G2. It is logical to suppose that during G1 the cell is preparing in some way for entry into DNA replication, but no such preparations have yet been specifically identified.

Enzymes important in DNA replication, such as DNA polymerase and those that catalyze synthesis of deoxynucleoside triphosphates, are not accumulated during G1; instead the numbers of these molecules crucial for DNA replication begin to increase around the start of the S period. The G1 period has been intensively studied, but aside from growth almost nothing is known about the molecular events in G 1 that contribute to the progression of a cell toward the S period.

The G1 period has a particular importance in that the rate of cell reproduction is controlled by regulation of transit through G1. Rapidly reproducing cells proceed through G1 quickly. With slower rates of reproduction, cells remain in Gi for longer times. Cells that temporarily or permanently stop proliferating, as happens in many tissues of multicellular organisms, remain in G1. Regulation is believed to be achieved by blocking a particular molecular event required to allow progression toward DNA replication. Hence it is important to determine how a cell progresses through the G1 period. More will be said later about arrest of cells in the G1 period and regulation of cell reproduction.

The initiation of DNA replication

The G1 period ends and the S period starts when the first replication units are triggered into replication. This is an important transition point in the cell cycle. Delaying or preventing a cell from reaching this transition is the means by which cell reproduction is regulated in multicellular organisms and in many unicellular eukaryotes. Entry of a cell into the S period commits the cell to another cell division.

How DNA replication is triggered is not known. Synthesis of enzymes that catalyze synthesis of deoxynucleoside triphosphates (dNTPs) and synthesis of DNA polymerase increase sharply at the start of the S period. However, neither the synthesis of these enzymes nor production of dNTPs initiates replication. Instead these are simply coordinated with initiation by unknown molecular mechanisms.

Initiation of replication is in some way linked to cell growth. For example, if growth is stopped by depriving the cell of an essential nutrient or by inhibiting protein synthesis with a drug, DNA replication is not initiated. Further, a variety of observations suggest that initiation is triggered when cells reach a critical size. On the other hand, the apparent association between cell size and the initiation of DNA replication can be easily disrupted, that is, small cells can be made to initiate replication.

The start of the S period is marked by the activation of transcription of a number of genes. Some of these encode enzymes concerned with DNA replication, and their activation leads to sharp increases in synthesis of enzymes, as just noted above. For example, the transcription of at least four genes encoding enzymes in the synthetic pathway for thymidine triphosphate (TTP) is activated. Also, transcription of genes encoding histones is turned on at or slightly before the start of S. Identification of the signal that coordinates the activation of the various genes with initiation of replication of DNA is an important problem.

Progression through the S period

Each DNA molecule (chromosome) in the eukaryotic nucleus contains many replicating units. A human diploid cell contains about 70,000 replicating units, or an average of about 1300 in the DNA molecule of each of the 46 chromosomes. The replicating units are not all initiated simultaneously at the start of the S period, but rather initiations are staggered throughout the S period. In a mammalian cell, replication units are on the average 30 μm long and each replicates in about 15

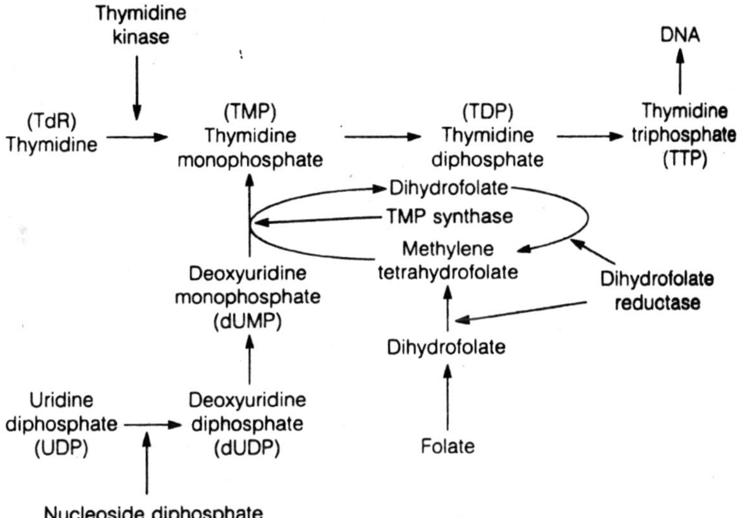

Figure 2.5: Enzyme pathways involved in the synthesis of thymidine triphosphate. Four enzymes (red) are turned on in the cell cycle at the transition from the G1 period to the S period and are turned off when DNA replication has been completed.

minutes. DNA replication in the nucleus is continuous over the eighthour S period, and therefore replication units must be initiated almost continuously over that interval.

The firing of replication units follows, at least to some degree, a regular sequential order. Replicating units that initiate at the start of one S period also initiate at the start of the subsequent S period. Replicating units that initiate late in one S period also initiate late in the next S period.

Order is also reflected by the replication of DNA in euchromatin in the first part of the S period and replication of DNA in heterochromatin in later S. This fits with two other experimental findings: (1) highly repetitious sequences, which are present in heterochromatin, generally replicate late in the S period, and (2) DNA that is actively transcribed is replicated in the first part of S, and DNA that is transcriptionally inactive (perhaps in heterochromatin) is replicated late in S.

Thus, so-called housekeeping genes, which encode proteins that are synthesized continuously in all cells, appear to replicate early. The timing of replication of genes that encode proteins that give a cell differentiated properties varies according to cell type. Thus, in cells that are differentiating into erythrocytes and still reproducing, genes encoding wand β-globins are replicated early. However the same genes are replicated late in cells in which transcription of the genes encoding α- and β-globins is turned off.

If protein synthesis is blocked in a cell in the S period, DNA replication quickly stops. This has been interpreted to mean either that histone synthesis is required for DNA replication to continue or that specific initiator proteins must be synthesized to trigger successive initiations in replication units. Beyond this, little is known about the control of initiation of the 70,000 replication units in a mammalian cell. The mechanism that controls the succession of initiations of replication units also provides that every replication unit is replicated and that normally no unit replicates more than once in a single S period.

The G2 period

The G2 period links the end of DNA replication with the initiation of mitosis, but not much is known about the molecular events that occur in this period. Inhibition of RNA or protein synthesis blocks transit through G2, so synthesis of new RNA and protein is necessary. Several proteins have been found to appear in G2 and then to disappear as soon as mitosis is over.

One of the proteins synthesized only in the G2 period is known as *maturation promoting factor (MPF)*. It brings about condensation of interphase chromosomes into the mitotic form. The existence of MPF was first demonstrated by a cell fusion experiment. When a mammalian cell in mitosis is fused with an interphase cell, the nuclear envelope of the interphase nucleus disappears and the chromosomes are caused to condense into the mitotic form, a phenomenon called *premature chromosome condensation*.

If the interphase cell used in this fusion experiment happens to be in the G1 period, then the prematurely condensed chromosomes appear containing a single chromatid each. Fusion of a mitotic cell with a cell in the S period produces chromosomes that have a fragmented or ragged appearance, consistent with what might be expected for chromosomes with DNA replication taking place at many points along their lengths. Fusion with a G2 cell produces condensed chromosomes containing two chromatids each, as expected.

Maturation promoting factor was later identified in mature oocytes of *Xenopus,* cells that remain arrested in metaphase of meiosis under the influence of MPF until fertilisation occurs.

Injection of MPF into cells of *Xenopus* embryos that had been experimentally arrested at the end of the S period causes dissolution of

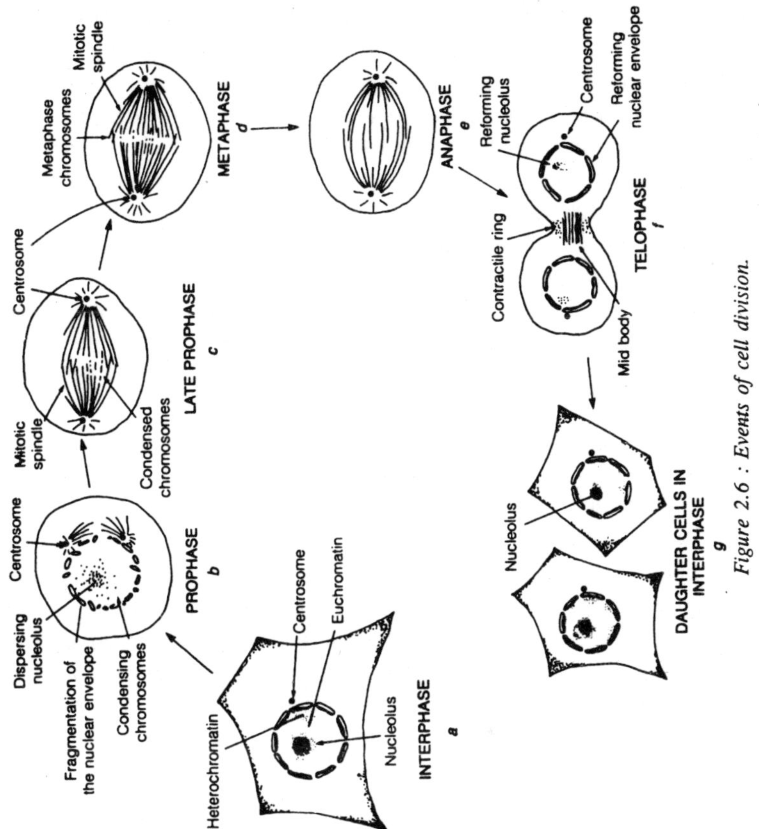

Figure 2.6 : Events of cell division.

the nuclear envelope and rapid mitotic condensation of the interphase chromosomes. MPF has also been identified in yeast cells in mitosis, in mammalian oocytes in meiosis, and in mammalian somatic cells. For example, injection of a semipurified MPF from mitotic mammalian cells into immature oocytes of *Xenopus* causes breakdown of the nuclear envelope and premature chromosome condensation.

Isolated interphase nuclei incubated in an MPFcontaining extract from *Xenopus* oocytes undergo breakdown of the nuclear envelope and chromosome condensation. This experimental system makes feasible a molecular analysis of MPF action and the mechanism controlling chromosome condensation. For example, within 15 minutes after addition of MPF, two of the major proteins (lamins) of the nuclear lamina (lamins A and C) become hyperphosphorylated.

This is followed by breakup of the nuclear lamina, presumably caused by hyperphosphorylation of lamins, and dissolution of the nuclear envelope 30 minutes later.

In the continued presence of MPF, cells remain in metaphase. An MPF-inactivating agent has been identified in oocytes of *Xenopus*. This agent appears in metaphase and inactivates MPF, which releases the cell from metaphase and allows it to complete mitosis (or meiosis) and enter interphase once again.

Blockage of cells in the G2 period occasionally occurs at a low frequency in somatic cells of plants and animals. For example, up to a few percent of the cells in the skin of mammals are arrested in the G2 period. When the skin is wounded, these cells are immediately released from the G2 block and enter mitosis. There are too few cells blocked in G2 to argue that they represent an important, rapid initial component in wound healing, and their significance remains unknown. They may occur because of a temporary failure of MPF to appear and trigger entry into mitosis.

The Events of Mitosis

Mitosis begins with prophase. This period of chromosome condensation is induced by MPF. It is the longest part of mitosis, lasting about 30 minutes in mammalian cells. Late in prophase the nuclear lamina and nuclear envelope disaggregate, and many nuclear components mix into the cytoplasm. With the breakdown of the lamina and nuclear envelope most of the nonhistone proteins of the nucleus and incompletely processed RNA transcripts are released to the cytoplasm.

During prophase, synthesis of mRNA and rRNA slows and finally stops (synthesis of tRNA continues at a reduced rate). In most kinds of cells the cessation of rRNA synthesis is accompanied by the complete dissolution of nucleoli. RNA synthesis stops presumably because the tight condensation of chromatin excludes access of transcription factors and RNA polymerase to the DNA. The situation is analogous to the effect resulting from packing of heterochromatin in interphase nuclei; transcription does not occur in heterochromatin.

In late prophase, about the time of breakdown of the nuclear envelope and dispersion of the lamina, the rate of protein synthesis in the cytoplasm begins to decrease. It falls to 30 percent of its *G2* value by the time metaphase occurs. The decrease in protein synthesis is not due to a decrease in the amount of mRNA, through normal turnover, since the life spans of almost all kinds of mRNA molecules in eukaryotes are much longer than the duration of mitosis. Some experiments suggest that protein synthesis slows because the ribosomes become incompetent

Changes at late prophase

Condensation of chromatin into mitotic chromosomes
Dissolution of the nuclear envelope
Dissolution of the nuclear lamina
Cessation of rRNA and mRNA synthesis
Dispersion of most nuclear proteins into the cytoplasm
Restructuring of the cytoskeleton and
Formation of the mitotic spindle
Decrease in protein synthesis

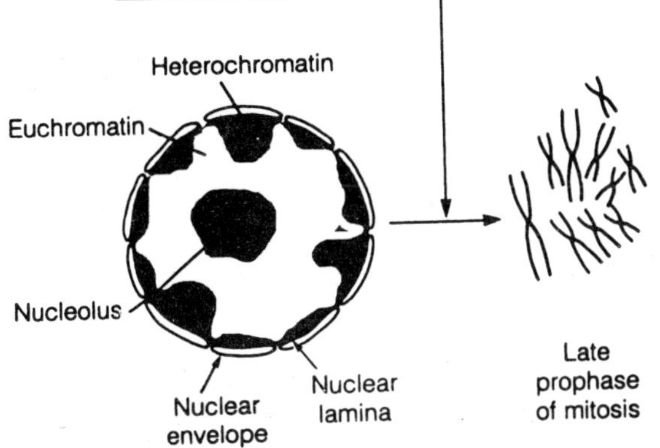

Interphase nucleus

Figure 2.7: Transit of a cell from interphase to mitosis is accompanied by dramatic changes in cell function and structure. All the changes are reversed at reentry into interphase after mitosis is over.

to carry out protein synthesis. This may be a result of proteins that become bound to ribosomes, perhaps nuclear proteins released when the nuclear envelope breaks down.

A transition period between prophase and metaphase, called *prometaphase*, is characterised by a shuffling motion of the condensed chromosomes in which they are brought into alignment in a plane in the center of the cell, *the metapha se plate*. How the central alignment is accomplished is not understood, but these events require the *mitotic spindle*.

The mitotic spindle forms in late prophase and prometaphase; it consists of microtubules and other components. The microtubules are assembled from tubulin molecules released by dissassembly of the cytoskeleton of the interphase cell. New tubulin synthesis occurs continuously in the cell but is not required for mitosis. A concomitant

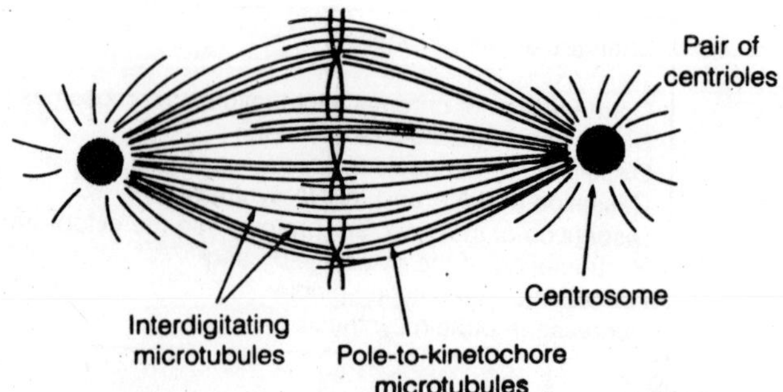

Figure 2.8: Schematic drawing of the mitotic spindle in an animal cell. Microtubules extend from the centrosomes toward the centrally aligned chromosomes. Two populations of microtubules are present. Centrosomal-kinetochore microtubules (red) and microtubules from opposite centrosomes that interdigitate in the central part of the spindle. A pair of centrioles is present in each centrosome.

of this disassembly in most kinds of animal cells and in some unicellular eukaryotes like amoebae growing on a surface is a drastic change from a flat irregular shape to a spherical form. In animal tissues, shape changes are less pronounced because of physical constraints provided by surrounding cells. The cell wall of plant cells and fungi prevents changes in cell shape during mitosis.

The microtubules of the spindle originate from opposite poles of the cell and interdigitate in the central region of the cell. An amorphous material occupies some of the space between microtubules. The two poles from which the microtubules emanate are called the *centrosomes*. The molecular nature of centrosomes has not been determined. Pure centrosomes have not yet been isolated in large enough quantities to do biochemical analyses.

The center of a centrosome in animal cells is occupied by a pair of centrioles. Plant cells and many unicellular eukaryotes, *Amoeba proteus* for example, lack centrioles in their centrosomes. Even in animal cells the centrioles appear not to contribute to the function of the centrosome.

Although cells of *Drosophila* normally have centrioles in their centrosomes, a culture of *Drosophila* cells has been obtained in which centrioles are absent and mitosis proceeds in the usual way. The presence of centrioles in centrosomes may simply represent a role of centrosomes in distributing centrioles equivalently to daughter cells. Centrioles serve as nucleating structures for assembly of microtubules in the formation of cilia and flagella.

Centrosomes act as nucleating centers for assembly of tubulin into microtubules in the formation of the mitotic spindle. Tubulin molecules are continuously added to the free (distal) ends of the microtubules, which grow toward the opposite centrosome. As the growing microtubules invade the central region of the cell in late prophase, some of them terminate by attachment to the chromosomes at their kinetochores.

Therefore, two types of microtubules are identifiable in the mitotic spindle: (1) interdigitating microtubules that grow out from a centrosome and continue through the central region of the cell, and (2) centrosome-to-kinetochore microtubules that connect kinetochores to centrosomes. Interdigitating microtubules form, disassemble, and reform continuously during mitosis. Centrosome-to-kinetochore microtubules are stabilised by their attachment to kinetochores; attachment appears to act as a cap for microtubules.

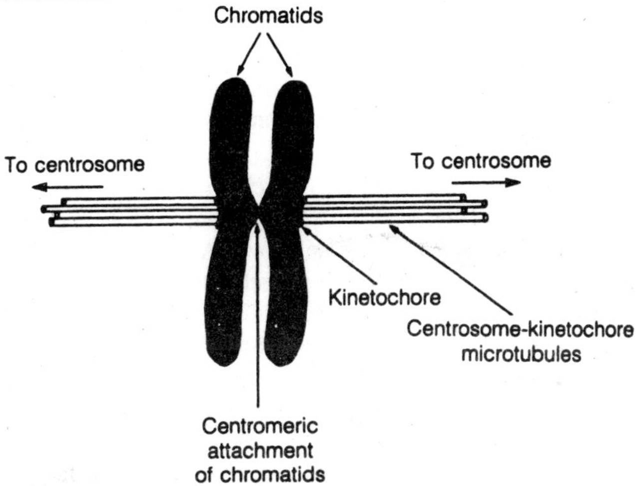

Figure 2.9: The plate-like kinetochore on a chromatid is composed of proteins and serves as an attachment site for microtutubles from a centrosome. Chromatids remain attached to each other at the centromere; detachment marks the beginning of anaphase.

Both chromatids in a chromosome have a kinetochore. The two kinetochores in a pair of chromatids face outward on opposite sides of the chromosome. The two chromatids remain attached to each other at the region on their inner sides from kinetochores. This attachment region plus the kinetochores together comprise the centromere.

Microtubules emanating from one centrosome attach to the kinetochore of one chromatid, and microtubules from the other centrosome attach to the kinetochore in the other chromatid. These attachments are essential for the back-and-forth motions of the chromosomes that eventually bring them into alignment in the metaphase plate. The completion of alignment defines arrival at metaphase. At this time, because of the microtubule attachments, one chromatid of each chromosome faces one centrosome and the other chromatid faces the opposite centrosome.

In somatic cells the chromosomes remain poised in mitotic metaphase for only a few minutes. However, during *meiosis* in many species of animals, chromosomes may remain in metaphase for months and even years. Mitotic metaphase ends and anaphase begins abruptly with the separation of the two chromatids in all the chromosomes simultaneously.

The nature of the connection between chromatids, how the connection is broken, and the signal that brings it about are not known. Anaphase is defined as the stage in which the two groups of daughter chromosomes move to opposite poles; it takes only a few minutes. Like the shuffling movement of condensed chromosomes during prometaphase, the movement of chromatids in anaphase is dependent on the centrosome-to-kinetochore microtubules.

The movement of chromatids, which have now become daughter chromosomes, has been intensively studied for decades; nevertheless, the molecular mechanism is still unknown. A protein, named *kinesin,* that appears to act as *a motor* to propel cytoplasmic particles along microtubules in nerve axons has been isolated from the mitotic spindle of animal cells. Its association with microtubules in the spindle suggests the intriguing possibility that kinesin may serve to connect the kinetochore to microtubules and may somehow provide the motor that propels chromosomes on microtubules in the spindle. An explanation of how daughter chromosomes move must also account for the fact that the centrosome-to-kinetochore microtubules shorten in correspondence with poleward movement of the daughter chromosomes.

When the two groups of daughter chromosomes reach the two poles of the cell, telophase begins. In telophase the chromosomes decondense, and a nuclear envelope and nuclear lamina reform around them. The envelope and lamina are reassembled from the components released by the earlier breakdown of the envelope and lamina in prophase.

Dephosphorylation of lamins accompanies their assembly into the lamina. Simultaneously, the several other changes that occurred in prophase are reversed. The nucleolus reforms around the nucleolar organizers of chromosomes, RNA synthesis resumes, and proteins released from the nucleus into the cytoplasm at the time of breakdown of the nuclear envelope in prophase quickly reaccumulate in the re-forming daughter nuclei. The mitotic spindle is disassembled, and the free tubulin molecules are reassembled into the cytoskeleton of the interphase cell. With completion of these telophase events, the nuclei enter the G 1 period of the next cell cycle.

Cytokinesis

Cytokinesis is the division of the body of the cell to form two daughter cells. Cytokinesis together with mitosis comprises cell division. The term mitosis is sometimes used incorrectly as a synonym for cell division; mitosis refers only to nuclear division.

Cytokinesis is coordinated in time and physical orientation with mitosis. It begins in late anaphase or early telophase, and division of the cell occurs in a plane at a right angle to the long axis of the spindle, midway between the separated groups of daughter chromosomes. In cells that lack rigid walls (animal cells, protozoa, some algae) cytokinesis differs from the comparable process in cells with rigid walls (plant cells, some algae, and most fungi). The two mechanisms of cytokinesis are described in the following two sections, typified by animal cells and plant cells.

Cytokinesis in animal cells

Cytokinesis in animal cells is a contractile process. The contractile mechanism is contained in a *contractile ring* located just inside the plasma membrane. The ring forms around the circumference of the cell in the region where the edge of the metaphase plate used to be. The ring consists predominantly of a bundle of microfilaments assembled from actin.

Also present in the contractile ring is myosin, one of the principal proteins that make up the contractile machinery in muscle cells. An interaction between actin and myosin, accompanied by hydrolysis of ATP to ADP and P; generates a contractile force that draws the contractile ring inward, forming a furrow in the cell surface and pinching the cell in two.

Part of the evidence that myosin is involved is the inhibition of cytokinesis (but not of mitosis) when antibody molecules that specifically bind to myosin are injected into single animal cells. To accomplish cytokinesis the contractile ring must be attached to the inside of the plasma membrane. As the ring contracts, it retains the same thickness, which means that actin and myosin are released from the ring as it draws inward. The contractile ring reduces the connection between the two daughter cells to a narrow bridge that contains a compressed bundle of interdigitating microtubules of the mitotic spindle. This compact bundle, called the *mid-body*, persists as the last remnant of the spindle and then disappears as the two cells finally break apart from each other. In animal cells the main part of cytokinesis occupies only a few minutes; the narrow bridge with its mid-body may persist for an hour or more in some kinds of cells. In some protozoa, such as the small amoeba *Acanthamoeba,* the contractile ring contracts extremely rapidly, and all of cytokinesis takes place in 40 to 50 seconds.

Even before cytokinesis is completed in animal cells, interphase cytoskeletons are assembling in the incipient daughter cells. As a consequence, the daughter cells flatten (if growing on a surface) and resume the amoeboid motion that stopped when the parental cell became spherical as it entered mitosis.

Cytokinesis in plant cells

Plant cells do not divide by means of a contractile ring. Instead they assemble plasma membranes and cell walls between the two daughter cells in the plane of the metaphase plate. The events of this assembly process have been defined in detail by observations in the light and electron microscopes.

When the daughter chromosomes have left the metaphase plate as a result of their anaphase movements, membranous vesicles derived from the Golgi complex accumulate in the plane where the metaphase plate used to be. The vesicles are drawn into the plane from both sides of the spindle, apparently by the action of microtubules.

The small vesicles fuse to form a large, flat vesicle that grows perpendicular to the spindle until it reaches the plasma membrane. The accumulated vesicular material is called the *phragmoplast (phragma* = enclosure, *plast* = forming). By their fusion the vesicles of the phragmoplast form a double plasma membrane. The two plasma membranes fuse peripherally with the main plasma membrane of the cell, thereby forming two separate daughter cells. Thereafter, a double cell wall is formed between the two plasma membranes by the daughter cells.

Although mitosis and cytokinesis usually occur together, this is not always the case. For example, in *Drosophila* eggs early development consists of many successive and synchronous mitoses in the fertilised egg without accompanying cytokinesis. This creates a single cytoplasmic mass with several thousand nuclei. At a precise point in development the nuclei migrate to positions close to the surface, and plasma membranes form to create simultaneously several thousand uninucleated cells.

Occasionally even in animal cells cytokinesis may fail to occur, creating binucleated cells.

Distribution of cytoplasmic components

In most instances cytokinesis divides a cell into roughly equal-sized daughters. Exceptions occur in early cleavage stages of embryos of many species, where cytokinesis produces large and small daughter cells in precisely defined patterns. Even in the usual case of equal division, cytokinesis often does not divide a cell into two exactly equal parts; one cell receives slightly more of the cytoplasm than the other.

This is of no major consequence since most cytoplasmic organelles like mitochondria and ribosomes are present in many copies and each cell always receives some of these. Failure of a daughter cell to receive at least one mitochondrion (and at least one chloroplast for plant and algal cells) would likely be lethal since these DNA-containing organelles form by fission of preexisting organelles and cannot form *de novo*. A cell lacking a Golgi complex or a contractile vacuole can form new ones *de nova*.

SYNCHRONISATION OF THE CELL CYCLE

The study of cell reproduction is based predominantly on measurements of the interrelated molecular events that occur during the cell cycle. Measurements on individual cells are sometimes possible, but usually they are difficult or impossible. Large populations of cells in which all the cells traverse the cycle in step with one another, that is, in synchrony, must be used for most kinds of molecular studies.

In some systems synchrony of the cell cycle in a population occurs naturally, a notable example being the cleavage of egg cells fertilised at the same time. Alternatively, a number of methods have been developed for obtaining synchronous masses of cells from populations in which the cells are originally completely out of synchrony with one another.

Synchronisation of Prokaryotic Cells

The simplest way to synchronize the cycles of bacterial cells like *E. coli* in a culture is to allow a population to enter the stationary phase of culture growth. As a result of depletion of one or another nutrient in the medium, cell growth is blocked, and all the cells arrest in the cell cycle between cell division and initiation of DNA replication. When the cells are removed from the exhausted medium, for example by sedimentation by centrifugation, and introduced into fresh medium,

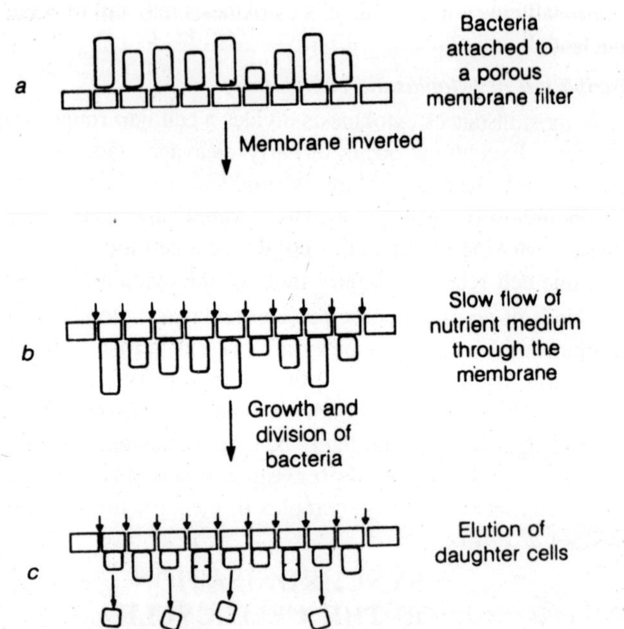

Figure 2.10: Synchronous bacteria obtained by the membrane filter method. (a) Bacteria are centrifuged onto a porous membrane filter, to which they remain attached. (b) Nutrient medium is allowed to flow slowly through the filter. (c) The cells grow and divide, and daughter cells are released into the medium and collected, for example, by centrifugation of the medium.

they all initiate DNA replication over an interval of many minutes and subsequently divide somewhat synchronously. The cells may retain a trace of synchrony for one to two cycles, but quickly become asynchronous because of variations in generation times of the individual cells.

The only method for obtaining highly synchronous populations of *E. coli* is to collect proliferating cells on a membrane filter. Nutrient medium is passed through the filter slowly and continuously so that the medium emerges from the side of the filter on which the bacteria have been collected. Initially, loosely attached cells are washed off, but subsequently only new daughter cells are released, and these can be collected as a synchronous population.

Using a series of filters, up to 5×10^8 synchronous cells per minute can be collected repeatedly. The increase in cell number for bacteria synchronised in this way is shown in Figure elsewhere in this chapter. By the first division after collection of synchronous cells, the degree of synchrony has seriously deteriorated, as indicated by the spread in the doubling of the cell population over a 15-minute interval.

Synchronous populations of cells make possible measurements of the transcription of particular genes, synthesis of particular proteins, or other activities in relation to progression of the cell cycle. Synchronous cells were used in the experiment described in other Chapter of this book for determining that DNA replication is initiated bidirectionally in the chromosome of *E. coli*.

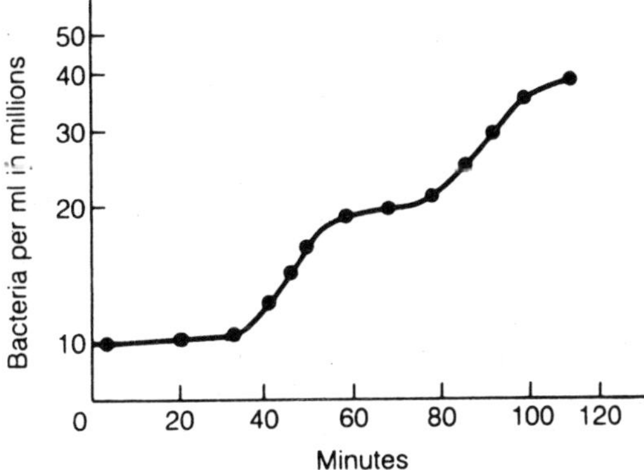

Figure 2.11: Cell increase in synchronous bacteria obtained by the membrane filter. The original high degree of synchrony at 0 time has deteriorated considerably by the first cell doubling and further decreased by the second cell doubling.

Synchronous Eukaryotic Cells

Many strategies have been developed for obtaining synchrony of the cell cycle in eukaryotes. These methods fall into three categories: natural synchrony, chemically induced synchronisation, and mechanical selection to obtain synchronous cells.

Natural synchrony

Fertilisation of an egg cell is followed in many species by extremely precise synchrony of division of the cells of the early embryo (*blastomeres*). In amphibians, synchrony lasts for more than ten cell cycles. These cycles are rapid and lack G1 and G2 periods. They become asynchronous suddenly at a particular cell division, for example, division 12 in *Xenopus* embryos, at which time the embryo contains 2^{12} (4096 cells). In the thirteenth cycle both G1 and G2 periods appear, and asynchrony is already apparent at the thirteenth division. Within a few more cycles the cells become completely asynchronous.

The divisions in embryos follow an exact time schedule, so groups of eggs fertilised at the same time undergo subsequent cell divisions with each other. Fertilisation of eggs therefore provides large numbers of cells reproducing with synchronous cycles. An example of the usefulness of such material is the study of the role of MPF in controlling initiation of mitotic condensation of chromosomes discussed earlier in the chapter.

Natural synchrony also occurs in the plasmodial slime molds. A *plasmodium* is a single large compartment of cytoplasm containing multiple nuclei. The slime mold *Physarum* grows as a flat cytoplasmic mass that can be many centimeters in diameter and contain tens of thousands of nuclei. As the plasmodium grows (without cytokinesis) all the nuclei proceed through G1, S, G2, and M in close to perfect synchrony. *Physarum* has been used extensively to study the nuclear events of the cell cycle.

Another type of natural synchrony occurs in organisms in which cell division is controlled by a biological timing mechanism known as a *circadian clock (circa* = about; *diem* = a day). In these organisms cell division in a population of cells (or population of unicellular microorganisms) occurs within an interval of a few hours at a particular time, often during the middle of the night.

The circadian clock, whose biochemical mechanism is completely unknown, is set to *local time* through a photoreceptor that transmits temporal information to the clock by sensing when it is day and when it is night. The clock is then coupled to the cell division process so that cells that have completed G2 and are ready to enter mitosis must wait until the clock reaches its permissive *temporal window* before dividing.

Synchronisation of cell division by circadian clocks is just one example of the widespread occurrence of circadian (daily) rhythms among eukaryotic organisms. Circadian rhythms have been observed in both unicellular and multicellular organisms in such functions as cell motility, phototaxis, photosynthesis, and bioluminescence (production of light by an organism), as well as in the levels of specific enzymes, substrates, and cofactors and in metabolic processes such as DNA replication, RNA and protein synthesis, fatty acid synthesis, and respiration.

The circadian clock serves to couple the physiology and biochemistry of the organism to fluctuations in the daily environmental cycle, particularly light and temperature. The adaptive value of the biological oscillation is clear in the case of rhythms such as photosynthesis and bioluminescence.

In many organisms there is a daily rhythm in the synthesis of components of the photosynthetic machinery that results in a rhythm in photosynthetic capacity with its maximum in the middle of the day. On the other hand rhythms in bioluminescence show a maximum in the middle of the night as a result, at least in part, of a circadian rhythm in the synthesis of luciferase, the key enzyme in the lightproducing chemical reaction in the cell. The adaptive significance of other rhythms such as cell division are not as obvious, although some ecologists have suggested that different species could occupy different *temporal niches* by having their cell cycles phased differentially throughout the day.

Chemically induced synchrony

Blocking cells at a single point in the cell cycle with an inhibitor or by deprivation of an essential nutrient induces synchrony by chemical means. An example was introduced earlier in the blocking of initiation of DNA replication with fluorodeoxyuridine (FUdR), an analog of thymidine. Fluorodeoxyuridine is converted by the cell to FUdR monophosphate (FUdRP) by thymidine kinase, just as thymidine is converted to thymidine monophosphate (TMP) by the same enzyme.

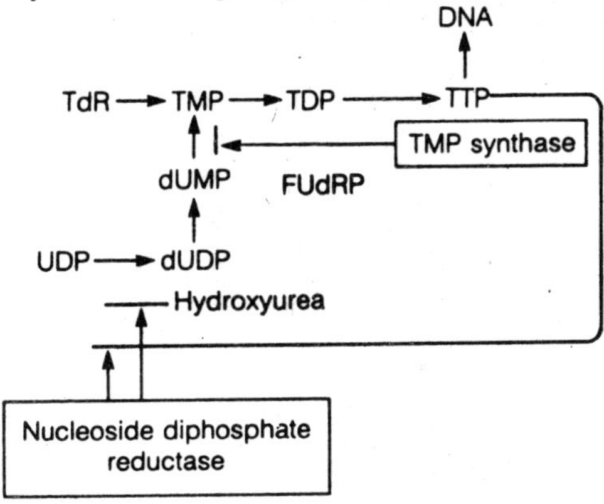

TdR = Thymidine
TMP = Thymidine monophosphate
TDP = Thymidine diphosphate
TTP = Thymidine triphosphate
FUdRP = Fluorodeoxyuridine monophosphate
dUMP = Deoxyuridine monophosphate
dUDP = Deoxyuridine diphosphate

Figure 2.12: Three ways to inhibit DNA synthesis by inhibiting the TTP biosynthetic pathway. Such inhibition of DNA synthesis is used to synchronize animal cells in culture.

FUdRP is a powerful inhibitor of the enzyme TMP synthase. The pathway for TMP synthesis is shown in detail in Figure elsewhere in this chapter because it is important for understanding the FUdR effect on the cell cycle of animal cells as well as the effects of two other inhibitors. The action of these inhibitors has additional significance because two of them are used as chemotherapeutic agents for combatting cancer by preventing DNA replication in cancer cells. TMP is phosphorylated to thymidine diphosphate (TDP) and then to thymidine triphosphate (TTP) in two successive steps. Thymidine triphosphate is required for replication of DNA. A block in its synthesis prevents cells from entering DNA replication and stops DNA replication of cells already in the S period when the block is imposed.

Thymidine monophosphate can be synthesized from two precursors, thymidine and deoxyuridine monophosphate (dUMP). Neither prokaryotes nor eukaryotes can synthesize thymidine from simpler precursors. Thymidine is only produced by the breakdown of TMP, for example, when DNA is degraded. For this reason thymidine kinase is called a "salvage" enzyme because it allows thymidine produced by degradation processes to be salvaged.

Ordinarily, thymidine is not available to cells in sufficient quantity to support DNA replication. Cells depend instead on synthesis of TMP from dUMP. All prokaryotes and eukaryotes produce dUMP by dephosphorylation of dUDP; dUDP is synthesized from uridine diphosphate (UDP), a process catalysed by nucleoside diphosphate reductase in which the ribose of UDP is converted to deoxyribose. In turn, UDP can be synthesized by most kinds of cells from simpler precursors.

There are several steps at which synthesis of TMP can be blocked by inhibitors, thereby preventing DNA replication and stopping cell reproduction. Inhibition at two of these steps is useful for synchronizing cells. Inhibition by FUdRP has already been mentioned. It binds tightly to the active site of TMP synthase and prevents the enzyme from catalyzing the addition of a methyl group to dUMP to form TMP.

As a result, cells reaching the end of the GI period are prevented from entering the S period. Cells already in the S period when FUdRP appears are stopped in S and are killed. We do not understand why cells die when blocked part way through the S period. Inhibition of TMP synthesis by FUdRP is easily overcome by adding thymidine to the medium. Thymidine is converted to TTP and all the cells blocked at the end of the G1 period by FUdRP initiate DNA replication synchronously.

The second useful inhibitor is hydroxyurea (HU). It inhibits the action of nucleoside diphosphate reductase, the enzyme that catalyzes both the synthesis of dUDP from UDP and synthesis of deoxycytidine diphosphate (dCDP) from cytidine diphosphate (CDP). Without dUDP and dCDP a cell cannot make TTP and deoxycytidine triphosphate (dCTP), and DNA cannot replicate. The result is accumulation of cells at the G1/S border. Washing of cells in fresh medium removes HU from the cells, relieves the inhibition, and allows cells to enter DNA replication in synchrony.

Thymidine can also be used to synchronize animal cells in culture in the following way. Several enzymes involved in production of deoxynucleoside triphosphates (dNTPs) are regulated such that levels of dNTPs adequate to support DNA replication without overproduction are maintained. Nucleoside diphosphate reductase is allosterically inhibited by TTP. When the level of TTP in the cell rises, it inhibits the reductase and prevents the synthesis of both dCTP and TTP. Thymidine added to the medium of a culture of animal cells is readily taken in and phosphorylated to TTP in an essentially unregulated manner.

If a high concentration of thymidine is added to medium (e.g., 2 mM), then the resulting high concentration of TTP in the cell inhibits nucleoside diphosphate reductase. This stops synthesis of TTP, which is of no consequence since the cell is already flooded with TTP. However, synthesis of dCTP is also inhibited, and without it DNA cannot replicate. The inhibition of the enzyme is not total, and cells continue to produce a trickle of dCTP.

As a result cells reaching the end of Gi enter the S period but replicate DNA extremely slowly and accumulate in the very first part of the S period. Cells already in S when a high concen-tration of thymidine is added to the medium continue progress through DNA replication, but at a very slow rate. Cells in G2, M, or GI are not affected in their rate of progress until they reach S.

If the cells are exposed to a high concentration of thymidine for an interval equal to the sum of the lengths of G2, M, and G 1, then all cells in these periods reach the S period and become tightly bunched in early S. Cells already in DNA replication remain scattered throughout the S period and progress very slowly. When the medium with the high thymidine concentration is replaced with normal medium, the intracellular concentration of TTP falls rapidly, relieving the inhibition of synthesis of dCTP.

Incubation of the cells in the fresh medium for a time equal to the length of the S period allows all the cells (theoretically 100 percent of the cells are in S) to complete replication. The cells originally scattered at various positions in S will now occupy positions in various parts of G2, D, and G 1. The large fraction of cells originally arrested in early S will reach the G2 period. Thus, all the cells in the culture are in G2, D, or G1. When a high concentration of thymidine is again added to the culture, the entire population of cells will accumulate in early S.

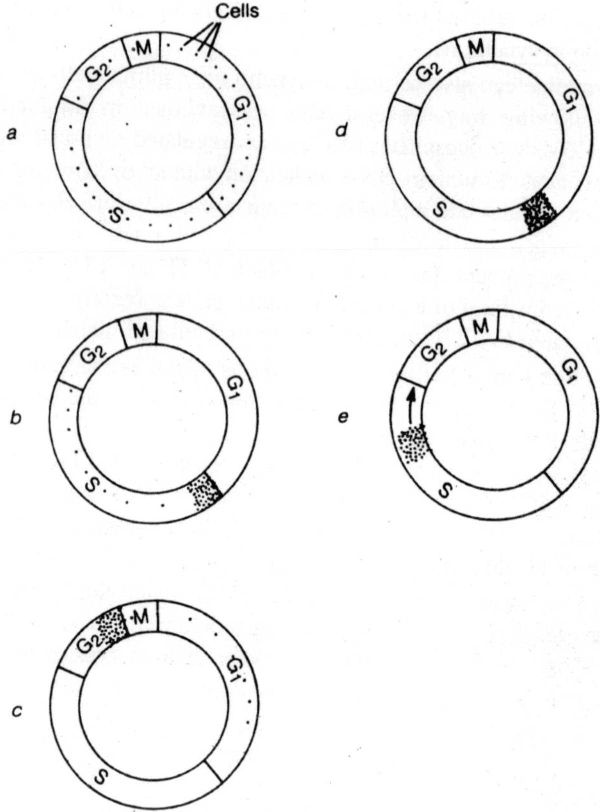

Figure 2.13: The double-thymidine-block method for synchronizing animal cells in culture. (a) Cells distributed throughout the cell cycle. (b) Treatment with 2 mM thymidine for 60 percent of a generation time accumulates all cells in S, with most in early S. (c) Release of the thymidine block allows all cells to progress into G2, M, and G1. (d) A second block with 2 mM thymidine accumulates all cells in early S. (e) Release of the second thymidine block allows the cells to proceed synchronously through the cycle.

This foregoing technique, called the *double-thymidine-block method* of synchronisation, makes it possible to impose an experimentally useful degree of cell-cycle synchrony via a normal metabolic regulatory mechanism instead of a toxic drug. The double-thymidine-block method avoids some of the uncertainty that attends the use of drugs. The latter may have undetected effects perturbing the very processes that the researcher is attempting to measure in the cell cycle.

Mechanical selection of synchronous cells

Synchronous animal cells can also be obtained by mechanical separation of cells according to their positions in the cell cycle. One such method is based on selection of animal cells in mitosis. Animal cells generally attach to the surfaces of culture vessels and form monolayers. In attaching they flatten and therefore have low profiles. During mitosis the cells disassemble microtubules (composed of tubulin) of the cytoskeleton and assemble mitotic spindles.

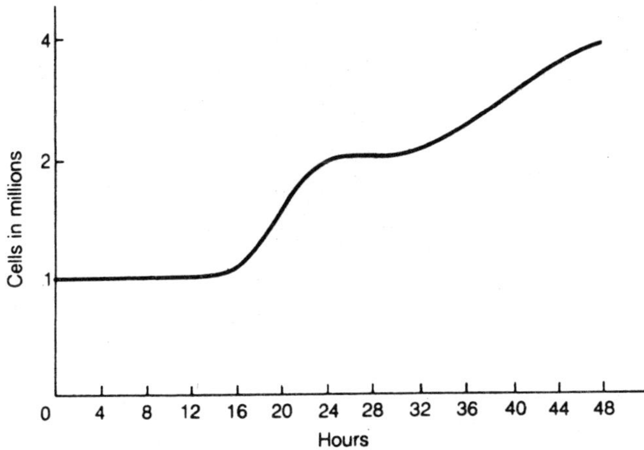

Figure 2.14 : Increase in cell number over two cell cycles in a population of synchronous mammalian cells obtained by mitotic selection. The experiment starts with one million new daughter cells, which proceed through interphase and divide into two million cells over an 8-hour interval.

The cells lose their flat shape, become spherical, and remain tethered to the surface only by thin strands of cytoplasm. With a spherical shape mitotic animal cells protrude up into the medium farther than interphase cells do and, in contrast to the flat interphase cells, are only loosely attached. Shaking the culture vessel causes the medium to flow across the monolayer of cells, and the mitotic cells are preferentially dislodged. The medium is removed from the culture, and the mitotic cells are collected by sedimentation in a centrifuge and cultured.

By this procedure a highly synchronous population of cells, initially all in prometaphase, metaphase, anaphase, and early telophase, is obtained that enters G1 synchronously and can be used for studies of the cell cycle. The original culture that yielded the mitotic cells can be incubated with fresh medium and shaken again, as more cells reach mitosis, to provide multiple synchronous populations.

In addition, the double-thymidine-block method can be used for initial synchronisation, and the mitotic selection method applied when the synchronised cells arrive at mitosis, a procedure that greatly increases the yield of mitotic cells.

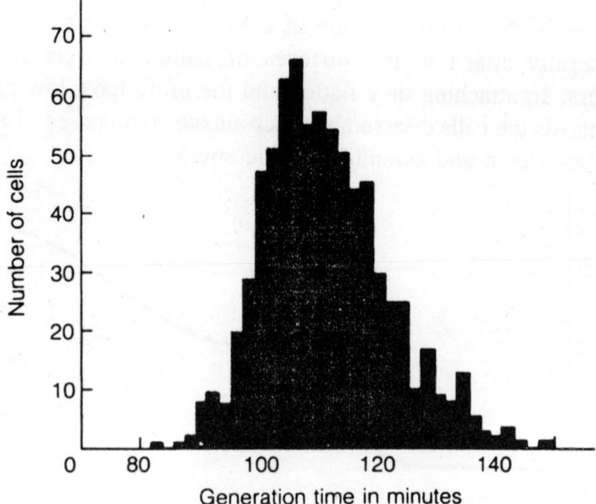

Figure 2.15 : Distribution of generation times for Tetrahymena.

Decay of Synchrony in Cell Populations

The best synchronisation methods depend on creating synchronous populations of cells at the G1/S border or at mitosis. The usefulness of these methods is limited by the rapid disappearance of synchrony in the population in subsequent cell cycles. Figure elsewhere in this chapter shows the increase in cell number in a culture initiated with mammalian cells selected at mitosis.

The cells begin to traverse the cycle in almost perfect synchrony. When they arrive at the next division, they are not so tightly in step with one another, doubling in number over an interval of several hours. The next doubling in cell number occurs with only a trace of the original synchrony, that is, the synchrony has decayed during traverse of the cycle. The third doubling in cell number occurs with no synchrony evident: increase in cell number is uniform with time and the curve is perfectly straight. Essentially the same result is obtained when cells are synchronised at the G1/S border by blocking DNA replication.

The decay in synchrony is essentially the same in every type of prokaryotic and eukaryotic cell. It occurs because of variation in the rate with which individual cells traverse the cell cycle.

A distribution of generation times is shown for several hundred individual *Tetrahymena* in Figure elsewhere in this chapter. All of these cells descended from a single starting cell, and hence they are genetically identical. The cells were grown in the same nutrient medium. Thus, the variation in generation times does not stem either from genetic difference or differences in nutrient availability or in other culture conditions. Generation times ranged from 84 to 149 minutes with an average of 111 minutes.

When the cells with the shortest and longest generation times were used to start two new clones, the distribution of generation times and the average generation time in the descendants of these two cells were the same as in the original clone in Figure elsewhere in this chapter. Hence, shorter or longer generation times are not inherited properties, but instead represent random fluctuations in the behaviour of individual cells from one cell cycle to another.

Random variation in generation times is due almost entirely to variation in the length of the G1 period. The combined lengths of the S, G2, and D periods is constant under a given set of culture conditions-in mammalian cells typically about 12 hours. The variation in the length of G i is easily measured by incorporation of ^{3}H-thymidine in a mammalian cell population synchronised by mitotic selection.

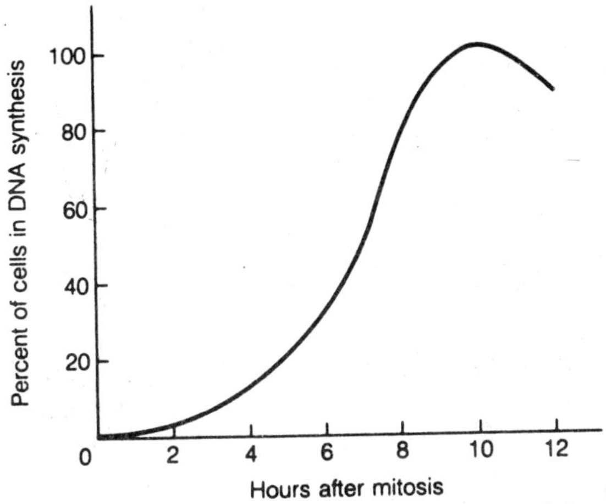

Figure 2.16 : Entry into DNA synthesis of cells in a synchronous population obtained by mitotic selection. The curve shows the variability of the G1 period of different cells.

The cells complete division, traverse G1, and begin to incorporate ^{3}H-thymidine when they reach the S period. Entry into the S period can be monitored by sampling the population at regular intervals and determining by autoradiography the percentage of cells that is labeled (*labeling index*). Figure elsewhere in this chapter shows the course of entry of cells into the S period.

The first cells reached S between one and two hours after mitosis.

The labeling index rose as more and more cells completed G 1 and reached nearly 100 percent at about 10 hours, by which time the first cells that reached S were now leaving S and entering the G2 period. Thus, the length of G1 varied from one to almost 10 hours in this experiment, with an average of about 6.5 hours. This variation in G1 length almost completely accounts for the variation in generation times among individual cells and the rapid decline in synchrony in an initially synchronous population of cells.

Variation in the length of G1 periods for individual cells occurs in a population in which the *average* G1 length and therefore the *average generation time* are constant. However, different kinds of cells in a multicellular organism usually have distinct and different average generation times. Such differences in average generation time are due to differences in the *average* length of the G1 periods in the different populations of cells.

These phenomena, variations in individual G1 periods in a homogeneous population of cells and differences in average G1 periods among distinct cell populations, should be carefully distinguished. Differences in average G1 periods represent regulation of reproductive rates of cell populations, as discussed below.

CELL-CYCLE GENES

Progress through the cycle is marked by changing activities in the cell, primarily centered around DNA replication and cell division. These activities require expression of a particular set of genes, called *cell-cycle genes*. For example, the initiation of DNA replication is accompanied by the turning on of genes encoding histones and enzymes for synthesis of deoxynucleoside triphosphates.

The initiation of chromosome condensation at the start of mitosis is dependent on expression of the gene encoding MPF. Other gene products are required for progress through other parts of the cycle, but little is known about these genes or the action of their products. One of the major unsolved questions is how DNA replication is initiated. Is it triggered by expression of a particular gene and, if so, what regulates expression of the gene?

More is known about cell-cycle genes in the budding yeast than in any other cell type. About 50 different genes have been identified by *conditional mutations*. A conditional mutation is one that affects the activity of the gene product under one culture condition but not another. Almost all conditional mutations discovered so far are temperature-sensitive, or *Ts mutations*. In most of these the mutation is expressed and results in a mutant phenotype at a temperature high in the normal temperature range for the cell. Such Ts mutations are said to be heat-sensitive.

A few Ts mutations are expressed as a mutant phenotype at a lower temperature and are normal at a higher temperature; these are cold-sensitive mutations. In most cases, Ts mutations are the result of an amino acid substitution that changes the stability of the three-dimensional configuration of the gene's protein product, making it nonfunctional above a certain temperature (in the case of a heat-sensitive mutation).

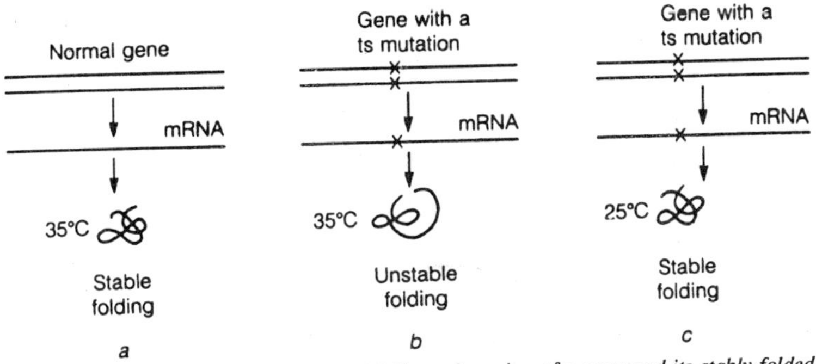

Figure 2.17 : Effect of a Ts mutation. (a) Normal version of a gene and its stably folded polypeptide product in a cell at 35°C. (b) A mutated version of the gene. The polypeptide product does not fold stably at 35°C. (c) The same mutated gene as in (b). At the lower temperature of 25°C the polypeptide folds stably.

A heat-sensitive mutation in a cell-cycle gene causes the cell to arrest at a point in the cycle at which the gene product is essential for progression. At a lower temperature the gene product is active, so the cell cycle can proceed normally.

The 50 cell-cycle genes identified in yeast by Ts mutations are crucial for progress at many points in the cycle. Some cell-cycle genes are required for initiation of DNA replication, some for progress through the S period, others function at various steps in mitosis, and still others are essential for cytokinesis. The analysis of the function of cell-cycle genes will eventually help elucidate how the cell is carried forward through its cycle.

Important examples of cell cycle genes identified by Ts mutations in yeast are six involved in control of entry into DNA replication. When any one of these genes suffers a Ts mutation, the cell cultured at 37°C is blocked at a point in the cycle located shortly before the start of the S period.

The same cell cultured at 23°C is not blocked. The point in the cycle identified by these six mutated genes is called *Start* in yeast cells.

When normal yeast cells at 37°C (or mutant yeast cells at 23°C) are starved for an essential nutrient, for example, phosphate, they arrest at Start. Similarly, prior to mating, yeast cells are arrested at Start by mating hormone.

The products encoded by the six Start genes identified by Ts mutation probably form a molecular switch that controls entry into the S period. One of these genes, called CDC-28 (CDC = cell division cycle), encodes a protein kinase, suggesting that operation of the switch involves phosphorylation of proteins.

REGULATION OF REPRODUCTION OF CULTURED CELLS

All normal prokaryotic and eukaryotic cells regulate their own reproduction in relation to nutrient conditions. When nutrients are insufficient to allow growth, the cells become blocked at a particular point in the cycle, usually in the G1 period.

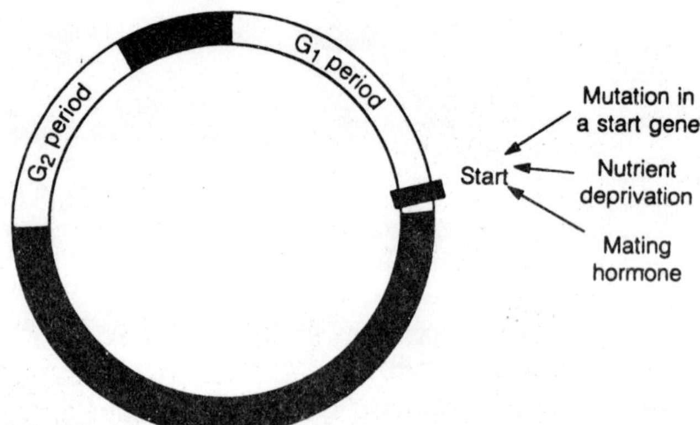

Figure 2.18: The Start switch in the cell cycle of the budding yeast is affected by mutation in a Start gene, nutrient deprivation, and mating hormone.

The cycle in a few organisms, such as *Amoeba proteus* and *Physarum,* lacks a G1 period, and these cells arrest in G2. Cells arrested in G1 (or G2) may survive for weeks, months, or even years in that state without loss of viability. Cells experimentally arrested in S with drugs quickly lose viability, usually within hours. Apparently, the normal regulatory mechanism ensures that the cell is arrested in a particular metabolic state in which it may remain without jeopardy, but what defines this state is unknown. Nor is it known why cells die when arrested in S for more than a short time.

In addition to nonspecific regulation by nutrient availability, the cell cycle may also be regulated by specific signals. Specific regulation occurs in some unicellular organisms, and an example in yeast is discussed in the next section.

Furthermore, evolution of specific regulation of cell reproduction was essential in the evolution of multicellular plants and animals, since unrestrained cell reproduction would be disasterous for a multicellular organism. What has been learned about regulation of cells in tissues during development of multicellular organisms and in adults is discussed in much of the remainder of this chapter.

Specific Regulation of the Cell Cycle in Yeast

Yeast, like most types of unicellular eukaryotes, can mate. Mating occurs between opposite mating types in a species. The mating type of a cell is determined by expression of a mating type gene. In the budding yeast cells are either mating type *a* or α.

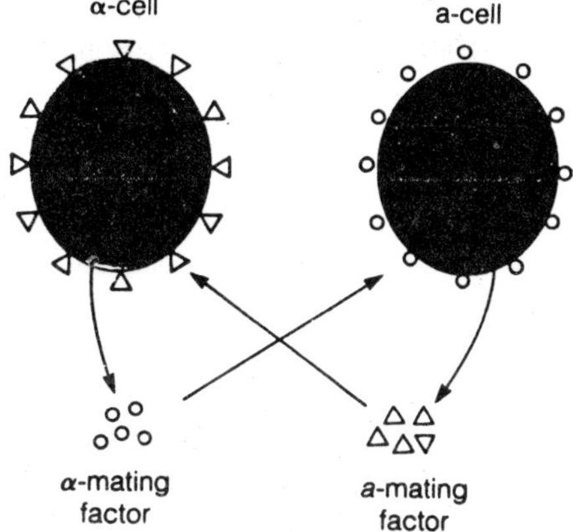

Figure 2.19 : Mating in budding type yeast. An a-mating type cell secretes α-mating factor, which binds to receptors on the membrane of an a-type cell. Reciprocally, an α-type cell secretes α-mating factor, which binds to receptors on the membrane of a a-type cell, Binding of mating factors arrests the cell in G1.

Yeast cells proliferate as haploid cells. In mating, an a mating type cell fuses with an a type, producing a diploid cell. The diploid cell can multiply if nutrient conditions are adequate or, if starved, undergoes meiosis and forms four haploid spores. The spores are metabolically inert, protected by a tough wall, and highly resistant to heat, toxic chemicals, and drying. Spores remain viable for years. Under favourable nutrient conditions they germinate and again proliferate as haploid cells until the next mating.

Mating of yeast cells occurs only between cells that are in the G1 period. The two mating types cause each other to arrest in G1 by secreting a small polypeptide, a mating hormone, that arrests cells of the opposite mating type at a specific point in G1 called Start. A 12amino-acid polypeptide, called a-factor, is secreted by a-type cells.

This polypeptide binds to specific receptors on the surface of a-type cells and causes them to arrest at Start in G1. The a-factor has no effect on a-type cells. Cells of a mating type secrete a polypeptide of ten amino acids-a factor-that causes atype cells to arrest at Start, but which has no effect on a-type cells.

Thus, the two mating types signal each other to arrest at Start, following which they mate by fusing to make a diploid cell. The fused cells cease to produce mating hormones and also lose sensitivity to them and are released from Start. The diploid cell proliferates or undergoes meiosis depending on nutrient conditions.

The mating phenomenon in yeast is a simple model of how one cell may specifically affect the cycle of another cell, arresting it at a specific point in G1. The Start point in yeast is positioned 10 to 20 minutes before the initiation of DNA replication and is an on/off switch that controls entry into the S period. The Start switch can be operated by a number of signals.

Presence of the a-type mating hormone or deficiency in any one of a variety of nutrients in the environment puts the switch in the off position. As long as nutritional conditions are adequate for cell growth and mating hormone is not present, the switch can be turned to *on,* and the cell completes its cycle.

Proliferation of Vertebrate Cells in Culture

Many kinds of normal cells from various tissues of animals proliferate in culture when supplied with amino acids, carbohydrates, vitamins, inorganic ions, and serum. Serum is blood from which all the cells and clotting factors have been removed. Serum provides several hormones and growth factors (discussed later in this chapter).

Although the study of cultured cells has uncovered important facts about regulation of cell reproduction, the environment provided to cells in culture vessels is different from what is provided in intact tissues in ways that dramatically affect regulation of the cell cycle. In tissues in the body, cell reproduction is precisely regulated by mechanisms that are still poorly understood.

The study of regulation in intact animals is difficult because there are so many influences on cell reproduction that complicate design and interpretation of experiments. These variables can be eliminated or controlled with cultured cells, but in cell cultures only remnants of normal control mechanisms continue to be present, and cells proliferate with much less restraint than in tissues. Development of culture methods that approximate conditions experienced by cells in intact tissues is a slow and difficult task.

Density-dependent inhibition of cell reproduction

When normal cells are placed in a culture, they attach to and flatten onto the surface inside the culture vessel. As the attached cells multiply, they form colonies that expand and fuse with one another until the entire surface of the culture vessel becomes covered with a complete, single layer of cells (a *confluent monolayer*). When the monolayer is complete, cell reproduction stops, even though the nutrient medium is still capable of supporting more cell proliferation.

The arrest of cell reproduction in the completed monolayer is called *contact inhibition of growth* or *density-dependent inhibition of growth* because the cells appear to exert an inhibition on one another when sufficiently crowded in a monolayer. The inhibited cells stop in the G1 period, where they may remain for many days without adverse effect. When the cells are removed from the culture and seeded into new culture vessels, they resume progress through the cell cycle.

Density-dependent inhibition observed in cultured cells is generally assumed to reflect one mechanism by which cell reproduction is regulated in intact tissues. How the inhibition in culture works is not known, but the following four observations provide a better picture of the phenomenon:

1. Before the monolayer becomes confluent, addition of ^{3}H-thymidine to a culture shows which cells are still reproducing in cell colonies. Figure elsewhere in this chapter is an autoradiograph of cells incubated with ^{3}H-thymidine for 24 hours. All the cells at the periphery of the colony have replicated DNA during the 24hour interval, and many have divided. None of the cells in the interior of the monolayer have replicated DNA; they are, in fact, arrested in the G1 period. Clearly, crowding of cells in the interior of the colony somehow causes an inhibition that is not present at the periphery of the colony.
2. When fresh serum is added to a complete monolayer of cells, a large fraction of the cells is released from G1 arrest and divides once before again becoming arrested in G1 in an even more crowded state. This release suggests that the cells have exhausted essential growth factors provided by the serum. However, when the medium is withdrawn from a culture with a complete cell monolayer in which all cells are arrested in

G 1 and that medium is seeded with a low density of cells, the new cells proliferate to a complete monolayer. This proliferation shows that arrest in the first culture could not have been due to depletion of growth factors in the medium.

3. When a monolayer is *wounded* by drawing a glass needle across it and scraping cells from the surface, the cells immediately adjacent to the free surface are released from G1, move into the unoccupied area, and proliferate until the wound has been repaired with a monolayer of new cells. This suggests but does not prove that relieving cell-cell contacts relieves the G 1 inhibition.
4. Plasma membranes purified from cells in a GI-arrested monolayer inhibit cell reproduction when added to a culture with cells still proliferating. Plasma membranes from proliferating cells do not have this inhibitory effect, which supports the hypothesis that the inhibition of one cell by another is mediated by inhibitors appearing in the plasma membrane as cells become crowded.

Anchorage dependence of normal cells

Normal cells in culture survive and proliferate only when attached to the glass or plastic surface of a culture vessel. Normal cells secrete a large glycoprotein called fibronectin that sticks to surfaces of both the cells and the culture vessel and promotes attachment of the cells to the culture vessel. Fibronectin is believed to play a role in cell-cell adhesion in tissues and in cell migration occurring during development. Normal cells prevented from settling and attaching to the surface of the culture vessel when the medium is gently stirred die within a day or two without dividing. The requirement for attachment is called *anchorage dependence*.

Serum growth factors

Serum is required for the proliferation of normal cells in culture. Typically, serum makes up 10 to 15 percent of the nutrient medium; the other 85 to 90 percent is a solution of nutrients. If reproducing cells are transferred to medium containing less serum, for example 5 percent, they stop proliferating by arresting in the G1 period.

Serum contains growth factors that are necessary for cell reproduction. These growth factors are proteins. Two principal growth factors that when acting together fulfill the growth factor requirements of many normal cells in culture are *epidermal growth factor (EGF) and platelet-derived growth factor (PDGF).* Platelets are formed in bone marrow and are released into the blood. They are packets of cytoplasm that pinch off from giant cells in the bone marrow called *megakaryocytes.* Platelets carry factors essential for clotting of blood. When a blood vessel is damaged, platelets lyse and release clotting factors, preventing bleeding from the wound. Platelet-derived growth factor is also released from the platelets and is probably important in stimulating cell reproduction needed for wound healing.

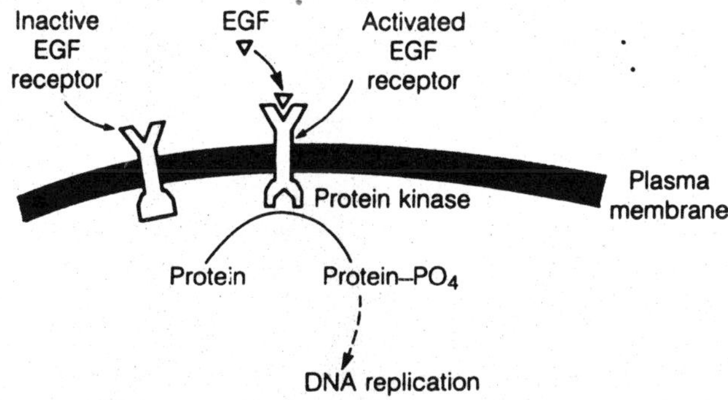

Figure 2.20 : The binding of epidermal growth factor (EGF) to the receptor at the outer surface of the membrane activates the protein kinase site on the cytoplasmic domain of the receptor.

Insulin in serum also acts as a growth factor for some kinds of normal cells. A variety of other growth factors are known that stimulate reproduction of specific cell types. For example, *erythropoietin* is a protein growth factor synthesised in the kidney. It stimulates reproduction of cells in the bone marrow that differentiate into red blood cells. *Interleukins* are growth factors that stimulate reproduction of certain kinds of white blood cells.

The effect of growth factors is mediated by means of receptors in the plasma membrane that specifically bind particular growth factors. For example, cultured normal cells that require EGF and PDGF have EGF and PDGF receptors. A receptor is a transmembrane glycoprotein that binds a signaling molecule like a growth factor or a hormone molecule to a binding site at the outer surface of the plasma membrane.

In the case of EGF receptor, binding of EGF at the cell surface alters the receptor protein and activates an enzymatic site in the part of the protein at the inner surface of the plasma membrane (the cytoplasmic side). The enzyme portion of the receptor is a protein kinase that phosphorylates particular proteins. The specific target proteins in the cytoplasm have not yet been identified, and the connection between the activated EGF receptor (receptor with EGF bound to it) and the regulation of cell reproduction is not understood in molecular terms.

Finite life span of normal cells in culture

Cells taken from normal tissue and put in culture have a limited life span. Typically, they are capable of 50 to 100 doublings (50 to 100 cell cycles) over a period of several months, and then die. The phenomenon is called *cell senescence*. Cells taken from a child will undergo more doublings than cells taken from an old adult; this might seem to support the idea that part of aging of a human may be due to loss of reproductive potential of cells. However, even the cells of an individual very old at the time of death still have the capacity to divide many times in culture.

Several theories have been proposed to explain why normal cells in culture are capable of only a limited number of reproductions. Most researchers believe that normal cells follow an intrinsic or intracellularly determined program (as opposed to extrinsic effects of the environment) that governs the number of reproductions of a cell. Such a program might consist either of the turning on or turning off of the transcription of certain genes, resulting in the closing down of reproduction. Whatever the mechanism, it works by arresting the cell permanently and irreversibly in the G1 period.

In the cultivation of cells the increase in cell number is logarithmic, and cells soon accumulate beyond the need for experiments and beyond what can practically be maintained. For example, during the first ten population doublings of cells newly explanted from a tissue, the cell number increases by 2^{10} (1000-fold). A convenient tactic is to store excess cells by freezing them in liquid nitrogen (-196°C) . They can be thawed and placed in culture whenever cells are needed, thus the more arduous procedure of explanting new cells from tissues can be avoided. Storage in the frozen state does not decrease or increase the reproductive potential of cells; those frozen after ten population doublings and stored for a year or more undergo the same total number of population doublings, when thawed, as cells never frozen.

G0 state

Cells arrested in G1 for whatever reason, for example nutrient deprivation, lack of growth factors, or density-dependent inhibition, undergo metabolic changes described as entry into quiescence or a *G zero* or *G0 state*. In G0, rates of RNA and protein synthesis are decreased.

One consequence of entry into G0 is a delay of some hours in the return of a cell to its reproductive cycle when the condition causing G1 arrest is reversed. Typically, 12 or more hours are required for cells to begin DNA replication after they are released from G0. Some of this time may correspond to part of the G1 period, but most of it is taken up by reversal of metabolic changes that occurred during transition into quiescence. For example, cells that cycle with an average G1 period of four hours require 12 or more hours to reach the S period after release from quiescence.

REGULATION OF CELL REPRODUCTION IN ORGANISMS

Observation of the behaviour of cells in culture provides some clues about regulation of reproduction of normal cells, but regulation is more complex in the tissues of intact organisms. First, regulation is different from one cell type to another. Reproductive rates range from very high in some tissues to zero in others. Second, reproductive rates for many cell types are flexible, for example, transiently increasing in response to injury to a tissue.

In general, three broad classes of cells are definable in animals by their rates of reproduction:

1. Some cell types reproduce continuously at high rates. Primary examples are cells in bone marrow and the lymphatic system that give rise to red and white blood cells, epithelial cells that line the alimentary canal, cells in the basal layer of the epidermis of skin, cells in hair follicles, male germ cells, and the epithelium of the uterus during the reproductive years in women.
2. Cells of some tissues reproduce at very low rates, for example, cells of liver, bone and other connective tissues, kidney, smooth muscle, pancreas, ovary, lung, and adrenal glands. Rates for some of these increase enormously in response to injury of a tissue, such as liver and bone, or in response to hormonal stimulation, such as smooth muscle of the uterus.
3. A few kinds of cells permanently lose the capacity for reproduction when they become differentiated. Examples are nerve cells, skeletal muscle cells, cardiac muscle cells, cells of the outer layers of epidermis, and mature red blood cells (which lack nuclei).

Differential rates of cell reproduction are achieved by arrest of cells in G1 for different lengths of time. Cells arrested for more than a few hours probably enter a G0 state comparable to that observed in cultured cells. Most bone marrow cells are only briefly arrested in G1. Liver cells in a rat divide on average less than once a year, and are almost constantly in G0. In the G0 state they perform the many differentiated functions of liver cells, such as breakdown of toxic substances like alcohol and other drugs that may be in the blood, production of bile, synthesis of certain blood proteins, and so forth. Nerve and skeletal muscle cells remain permanently in G0.

Stimulation of Cell Reproduction by Hormones and Growth Factors

Differential rates of cell reproduction are achieved by a variety of mechanisms. One of the better understood mechanisms is stimulation of cell reproduction by hormones and growth factors. For example, as described earlier, release of PDGF by platelets in the vicinity of a wound stimulates reproduction of cells as a part of wound repair.

The rate of reproduction of bone marrow cells that differentiate into erythrocytes is stimulated by the hormone erythropoietin, which is synthesized by kidney cells at a rate determined by the O_2 level of the blood. When the O_2 level falls, the production of erythropoietin increases in the kidney, which causes an increased rate of erythrocyte production in the bone marrow.

The increase in erythrocytes raises the capacity of blood to carry O_2. This regulatory mechanism not only maintains a rate of production of erythrocytes adequate to replace erythrocytes that wear out and die; the mechanism also steps up production of erythrocytes in response to loss of blood through bleeding and in response to the lower amount of O_2 in the atmosphere at a high altitude.

Reproduction of epithelial cells in the uterus is at least partially regulated by estrogen. Reproduction of male germ cells is at least partially regulated by the male hormone testosterone.

Regulation of Cell Reproduction by Inhibitors

In addition to stimulation of reproduction of specific target cells by growth factors and hormones, cell reproduction is also controlled by inhibitors.

An inhibitor of epithelial cells in culture

One identified inhibitor is a small protein molecule produced by epithelial cells cultured from monkey kidney. The inhibitor, known as TGF-β, is effective at an extremely low concentration (ng per ml of medium) and blocks the reproduction of epithelial cells that produce it, but does not block fibroblasts. It is presumed that the inhibitor is part of the mechanism by which reproduction of cells is controlled in the intact animal.

The blocking of reproduction of cells by an inhibitor they themselves synthesize is a form of *autoregulation*. Autoregulation of reproduction of liver cells is one of the better understood examples of this mechanism.

Autoregulation of liver cell reproduction

Practically all the cells in an adult mammalian liver are in the G0 state. The cell death rate in liver is very low and the rate of reproduction of liver cells is also correspondingly low. When part of the liver is removed by surgery, the remaining liver cells are triggered out of G0, reproduce, and replace the cells removed by surgery. Up to 75 percent of the liver can be removed from a rat, and the remaining 25 percent will reproduce to regenerate a full-sized liver in just a few days.

3

GERMINAL CELL REPRODUCTION

Mitosis is a device of increasing the number of cells. During asexual reproduction it is the means of multiplying the individual. Therefore, in asexually reproducing organisms, set of characters are transferred unchanged from parents to offspring, except of course accidents. There is uniparental inheritance due to mitosis. But in sexually reproducing organisms, gametes are produced as a result of meiosis. Two gametes unite to give rise to a zygote which divides mitotically to from the individual. More often than gametes from different parents unite to form a zygote, thus ensuring biparental inheritance.

This results into a blending of characters from two parents into their offspring and ensures variation which is important for evolution. Gametes contain half the number of chromosomes, as compared to somatic cells. Union of two gametes restores the somatic number of chromosomes in the zygote. Gametic union results into the mixing of cytoplasms of two cells which is followed by the fusion of the nuclei.

During meiosis or at the time of gamete formation, therefore, halving of the number of chromosomes or the amount of DNA is achieved by two successive divisions of cytoplasm and nuclei (meiosis I and meiosis II) accompanied by only one replication of chromosomes. For example, each somatic cells of man has 46 chromosomes. Somatic cells are derived from the zygote due to mitotic divisions.

Zygote is the result of a fusion of two gametes, each of which contain only 23 chromosomes. Both the male and female gametes contain the same amount of DNA and equal number of morphologically similar chromosomes. The zygote and somatic cells derived therefrom, therefore, contain 23 pairs of homologous chromosomes, one of each pair derived from the male parent through the male gamete and the other from the female through the female gamete.

The members of each pair are homologous to each other and nonhomologous to the other chromosomes. In some organisms, the chromosome complements of male and female cells are slightly different and this forms the basis of sex differentiation. Those cells which possess two sets of chromosomes, are called *diploid* and they have *2n* or diploid or unreduced or zygotic number of chromosomes. Gamets, on the other have *n*, or *haploid* or reduced or gametic number of chromosomes.

In sexual reproducing organisms the diploid and haploid phases of the life cycle alternate and this phenomenon in called *alternation* of *generation*. In plants, there is an alternation of sporophytic *(2n)* and gametophytic *(n)* generation. Reduction division heralds the beginning of gametophytic generation and fusion of gametes that of the sporaphytic phase. Thus, meiosis can be studied at the time of gamete formation which occurs in specialised organs i.e., the embryo sac and the pollen sac.

A single cell undergoes meiosis in the embryo but in the pollen sac a large number of spore mother cells divide meiotically, almost synchronously, to produce haploid pollen grains. It is, therefore, easier to study meiosis is anothers, similarly in testes in animals. Meiosis consists of two divisions which take place one after the other, during which the number of chromosomes is halved.

The two divisions are known as the first meiotic division or heterotypic division and the second meiotic division or homotypic division. The first meiotic division has the same stages as mitosis. These stages are known as prophase I, prometaphase I, metaphase 1, anaphase I and telopbase I. Except for prophase, the other stages are similar to those of mitosis. Following telophase I is a period of interkinesis or rest. The second meiotic division consists of prophase 11, metaphase II, anaphase II and telophase 11.

HETEROTYPIC OR 1ST MEIOTIC DIVISION

This is the first half of meiosis and is true reductional in nature. Here the diploid cell divides into two haploid daughter cell. As stated already that the phases of this division one called Prophase I, Metaphase I, Anaphase I and Telophase I.

Prophase I

It is the longest stage and its duration makes from a few hours to a few days. The nucleus starts increasing in volume and this increase is much than the mitotic increase to some extent. This is due to an increase in hydration.

Nobel and *Ruttle* have called this stage as the *premelotic spiral prophase*. Prophase I is Further divided into six substages.

(*a*) Pre-leptotene | (*d*) Pachytene
(*b*) Leptotene | (*e*) Diplotene
(*c*) Zygotene | (*f*) Diakinesis

- (*a*) Pre-leptotene
- (*b*) Leptotene
- (*c*) Zygotene
- (*d*) Pachytene
- (*e*) Diplotene
- (*f*) Diakinesis

(a) Preleptotene

As noted previously, replication of DNA has already taken place during interphase. The chromosomes as the beginning of meiosis are, therefore, not single but double strands. The beginning of meiotic prophase is marked by an increase in nuclear volume. During preleptotene the chromosome are very thin and cannot be easily seen, except sometimes for the sex chromosomes.

(b) Leptotene ("slender thread") or Leptonema

During this stage the chromosomes become more distinct, and their double nature is seen in many organisms. Leptotene extends from the time single unpaired chromosomes are first seen to the time when they begin to undergo homologous pairing. Under the light microscope the chromosomes appear as slender threads bearing a series of granule-like structure called *chromomeres*.

Whole mount electron microscope preparations show that leptotene chromosomes have an axial filament to which chromatin fibres are attached as a series of lateral loops. In certain regions these loops appear in bunches, and from the chromomeres of light microscopy. The axial filaments are apparently composed of two independent protein fibrils, one for each chromatid.

They later become transformed into the lateral elements of the synaptonemal complex. In the leptotene cells of the quail there is evidence of premeiotic pairing. Two homologous chromosomes become apposed to each other, chromomere for chromomore, eve though no synaptonemal complex is formed.

In leptotene cell' there is a tendency for the ends of homologous chromosome to be associated with each other at the nuclear membranf where they are attached. Leptotene chromosomes may be irregularly arranged, or may be polarised toward the oentrioles forming a *'bouquet'*.

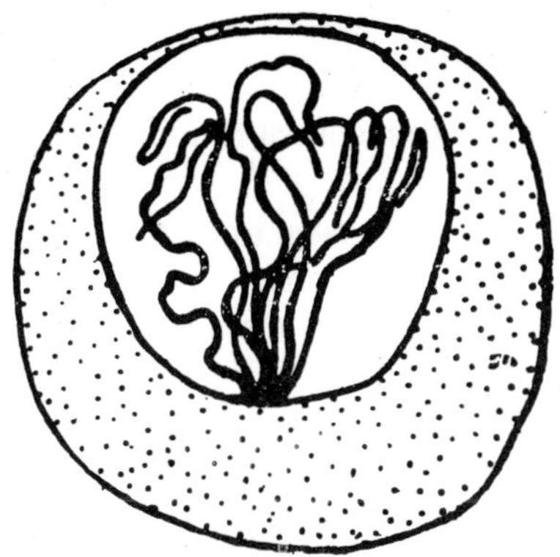

Figure 3.1 : Leptotene chromosomes arranged in a bouquet'.

Electron microscopic studies have shown that bouquet formation results when a group of chromosomes is attached close together on the nuclear membrane.

During leptotene the cytoplasm has many polyribosomes, but endoplasmic vesicles are few. In maize *(Zea mays)* the nucleus increases in size during leptotene, reaching its maximum size during pachytene. This increase in size is correlated with synthesis of RNA and protein.

(c) Zygotene ("mating thread") or Zygonema

The chromosomes becomes much shorter and thick and number of coils is also increased. Homologous chromosomes begin to pair by force of attraction. In synapsis or pairing, two homologous chromosomes come close to each other precisely chromomere to chromomere.

The pairing starts from one point and proceeds onward in a zipper like fashion. Out of the two synaptic chromosomes, one is of paternal origin while the other is maternal. These are called *bivalents*. Since each chromosome is made of two chromatids, each pair of chromosomes consists of four chromatids thus representing *'tetrad'* of chromatids at this stage.

The pairing of zygonemal stage is completed in three different ways, as follows

(i) Proterminal synapsis

The two homologous chromosomes start pairing at the terminals, which gradually then progresses towards the centromere.

(ii) Procentric synapsis

The pairing starts near the centromere and progresses onwards on either side.

(iii) Random synapsis

The pairing may be at random at many points which then finally fuse together.

Two views have been advocated to explain the mode of initiation of synapsis. *Darlington* proposed precocity theory which explains that the chromosomes undergo synapsis due to singleness, but this theory does not explain the extra synthesis of DNA and duplication at leptotene stage.

Another view is proposed by *Sax* which is referred as the retardation theory. It explains that pairing takes place due to the retardation or cessation of metabolic activities of the cell. During zygonema, the nucleus remains intact and the centriole moves apart initiating the aster formation.

(d) Pachytene "thick thread" or pachfynema

It generally begins when the pairing of chromosomes ends. It represents one of the longer substages of prophase I and is characterised by the following important events. When the pairing has completed, the aposition of homologues becomes so tight that it is difficult to identify two separate chromosomes.

Within the major coils, and at right angles to them smaller minor coils appear. At the height of pachytene, the pairing chromosomes twist about one another as relational coils and each soon longitudinally splits if no splitting had occurred. This coiling is further complicated by a coiling of two chromatids of each chromosome around each other. This coiling put the chromosomes under considerable strain.

This means that at this stage, each pacytence element consists of four chromatids. The chromatids of each homologue are called *sister chromatids*. Formerly this element of four chromatids was called a *tetrad*-a name which has now been replaced by *bivalent*. When the chromosomes have longitudinally splitted, transverse breaks occur at the same level in two of the adjacent homologous chromatids. Soon both segments of the chromatids interchange and fuse together with those of homologues.

In this process which is called *crossing over*, portions of the two chromatids with their genes are interchanged and two remain intact. The new chromatids thus interchanged will be mixed. After this interchange, the parts begin to separate, repalling each other. However, this separation is not at all complete since the homologous chromosomes remain joined at their point of interchange. This point where the two are joined is called the *chiasma* (pl. chiasmata). This chiasma is an immediate consequence of crossing over and are found in all plants and animal with few exceptions (e.g., males and females silk worm).

The number of chiasmata is variable but for one bivalent atleast one chiasma is found. Bivalents with 2, 3 and 4 chiasmata are rare. The highest number of chiasmata has been observed in the long chromosome of the broad bean, *Vicia faba,* where there are probably 12 chiasmata in a bivalent. The average number of chiasmata is known as *chiasma frequency.*

This number seems to depend upon the length of the chromosome. Short acrocentric chromosomes have a greater number of chiasmata than metacentric chromosomes. In small chromosomes, chiasmata are localised at the distal ends. Nucleus remains as such, being somewhat larger than the proceeding substage.

Cytomixis

It is a botanical term which refers to the passage of chromatin from one microsporocyte (pollen mother cell) to another during early prophase. This phenomenon is so widespread in the normal species of hybrids that one wonders if it is a general phenomenon.

(e) Diplotene or Diplonema

The separations of homologous chromosomes initiates diplotene. The synaptic forces of attraction between them lapse due to breakage at one or more point, so that thehomologous chromosomes uncoil and start separating. But the separation is nonetheless incomplete since the homologues are in, contact at one or more points where the crossing over has already taken place.

These points of contact are known as chiasmata (sing-chiasma ; meaning cross) which present crosshaped appearance. The chiasma is the morphological equvalent of genetic crossing over. Their number and position varies in the same pair of chromosomes and in different cells of the same individual.

By the end of diplotene, the chiasmata begin to move along the length of chromosomes from the centromere towards the end. This displacement of chiasmata is termed as terminalisation. When the terminalisation of chiasmata is complete the homologous chromosomes are kept in contact by terminal chiasmata. The degree of terminalisation is generally expressedd as a coefficient of terminalisation (T).

$$T = \frac{\text{Number of terminal chiasmata}}{\text{Total number of chiasmata}}$$

The average number of chiasmata in bivalents is known as frequency

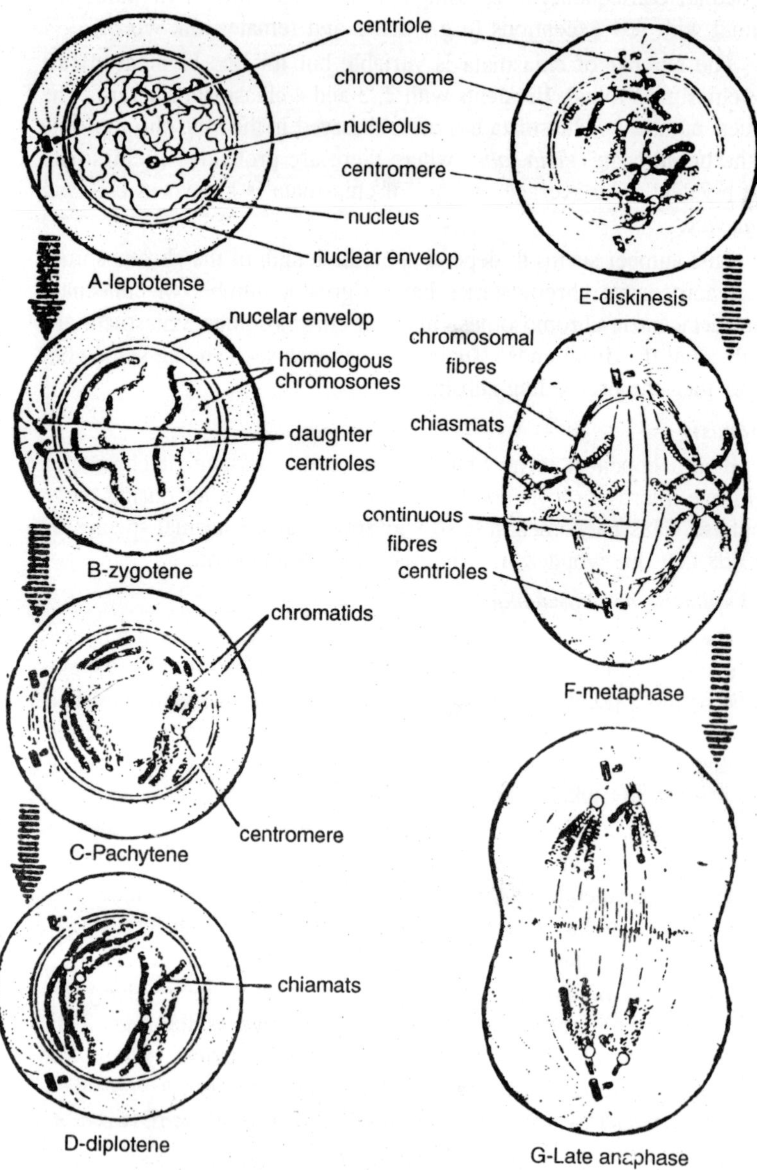

Figure 3.2: Different stages of first meiotic division.

of chiasmata (fq).

$$\text{Frequency (fq)} = \frac{\text{Total number of chiasmata}}{\text{Total number of bivalents}}$$

According to *Darlingtion* two types of repelling forces operate on the chromosomes at diplotene. One of the forces is electronegatively charged and operates on the surface of the chromosome throughout its length and the other with electropositive charge is localised on the centromere. The former controls the repulsion of the chromosomes and the later causes distal movement of the chiasmata.

Diakinesis

At this stage the chromosomes are best seen as very short and thick

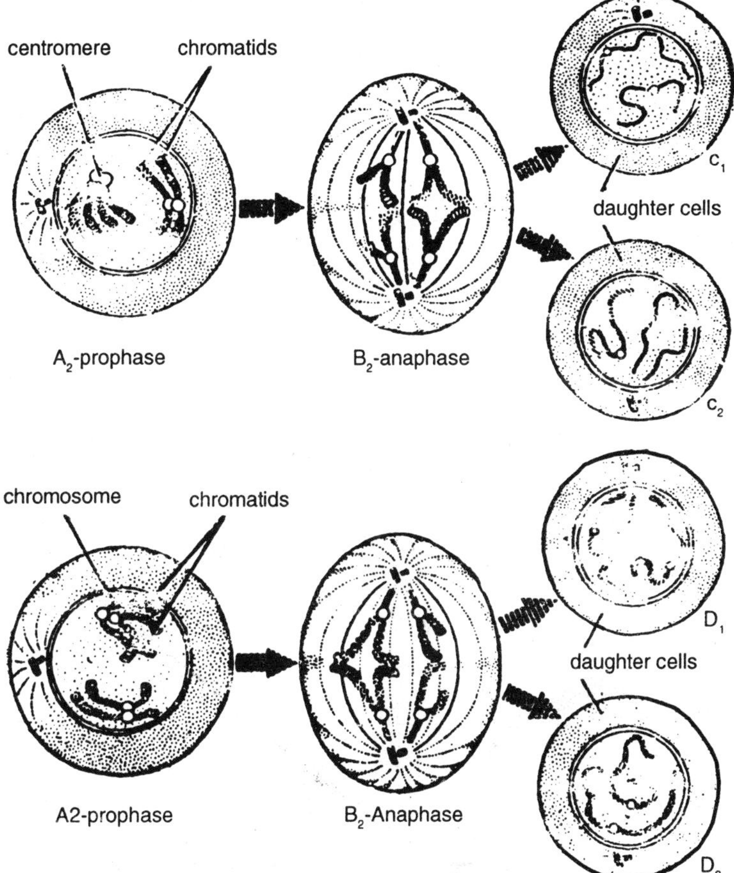

Figure 3.3: Different stages of second meiotic division.

bivalents. The contraction is nearly maximum and the chromosomal pairs are well spread throughout the cell as though by mutual repulsion. The chromatids of each chromosome are very close to each other and their identity is not distinguishable.

Terminalisation may be completed in the substage. The bivalents migrate to the periphery of the nucleus. The nucleus generally disappears during this stage but may persist, in reduced size until anaphase. The nuclear membrane disrupts and eventually disappears thus all the chromosomes are set free in the cytoplasm. The spindle fibres are organised with an establishment of the pole.

Prometaphase

Spiralisation reaches its virtual maximum in this stage ; the nuclear membrane disappears, and the chromosomes become arranged at the equator of the cell to begin the metaphase. Nuclear membrane disappears and spindle and astral rays are formed during this stage.

Metaphase I

Just as in mitosis, breakdown of the nuclear envelope provides a convenient reference point for the transition from prophase I to metaphase I in most higher eukaryotes. The spindle invades the position formerly occupied by the nucleus, and the tetrads, left scattered by the breakdown of the nuclear envelope, move to the spindle midpoint.

Except for the pairing of homologous chromosomes into tetrads, meiosis at this point closely matches the transition from prophase to metaphase in mitosis.

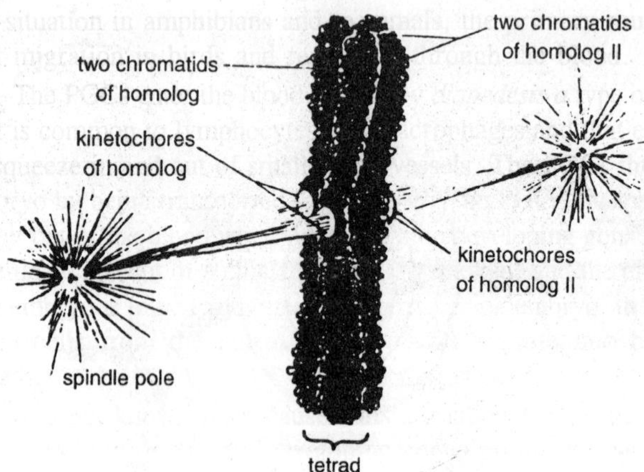

Figure 3.4: The spindle connections made by the chromosomes of a tetrad at metaphase I of meiosis.

The first major divergence from the mitotic pattern- and in fact, the distinctive event of the meiotic divisions following prophase I-occurs as the tetrads make their attachments to spirdle microtubules. An electron microscope study by P. Luykx demonstrated conclusively that each homologous chromosome has two kinetochores, one for each of its two chromatids.

As a consequence, four kinetochores are present in each tetrad. But even though two kinetochores are present in each bomolog, Luykx's micrographs show that both of these make connections to the *same* pole of the spindle at metaphase I. The other homolog of a pair makes attachments to spindle microtubules leading to the opposite pole. Thus, the two kinetochores of a homolog act as a functional unit at metaphase I of meiosis.

Luykx found that the two kinetochores of a homolog may initially make connections to opposite spindle poles early in metaphase I, but the incorrect attachments are broken and the correct attachments made by the completion of metaphase I. Similar results were also obtained by R. B. Nicklas, who detached the tetrads from a metaphase spindle by micromanipulation and reoriented them at various angles to the piles.

Although incorrect microtubule connections were initially made by a large proportion of the disturbed tetrads, these were quickly corrected so that, in each case, the two kinetochores of each homolog again made connections to only one pole. How the kinetochores sort out the correct connections at metaphase I of meiosis is one of the many unsolved mysteries of the meiotic mechanism.

Anaphase I

Separation of the homologous pairs and initiation of poleward movement by the homologs mark the beginning of anaphase I. Because of the opposite connections made by the kinetochores, the anaphase movement separates the two cbromosomes of each homologous pair and delivers them to opposite spindle poles.

At the completion of anaphase I, as a consequence, the poles receive the haploid number of chromosomes. But because each chromosome is still double and contains two chromatids, the 2 × amount of DNA is present at each pole.

The different homologous pairs separate independently from each other, so that any combination of chromosomes of maternal and paternal origin may be delivered to a spindle pole. The random combinations delivered to the poles contribute to the total genetic variability of the products of meiosis. In a human, for example, with 23 pairs of chromosomes, 2^{23} combinations of maternal and paternal chromosomes are possible. Even without recombination, therefore, the probability that tvvo children of the same parents will receive the same combination of maternal and paternal chromosomes would be one chance out of $(2^{23})^2$, or one out of 70,000,000,000,000!

The further variability introduced by recombination, which mixes chromatid segments randomly between maternal and paternal chromatids, makes it practically impossible for an individual to produce genetically identical offspring (except for identical twins).

Telophase I and Interkinesis

Well-defined telophase I and interkinesis stages do not always intercede between anaphase I and the second meiotic division. All possible gradations are found in nature, from organisms with no detectable decondensation of the chromosomes at telophase I and interkinesis to species with almost complete reversion to an interphaselike state. In the latter organisms, the chromosomes decondense and nuclear envelopes temporarily surround the polar masses of chromatin.

However, in most species, telophase I and interkinesis are transitory stages between the first and second meiotic divisions in which the meiotic cells pause only briefly before entering prophase II of meiosis. No DNA replication occurs at this time in any known organism. During this period, the single metaphase spindle is reorganised into two spindles that form in the regions of the telophase I spindle poles. Centrioles and asters, if present, also divide at this time, placing a single centriole at each pole of the two spindles. These events complete the cellular rearrangements leading to the second meiotic division.

HOMOTYPIC OR 2ND MEIOTIC DIVISION

Bofore the start of this division, most of the activities of heterotypic division cease for some time. This resting stage is termed as the interkinesis. 2nd Meiotic division involves following steps:

Prophase II

This stage superficially resembles that of the mitosis prophase with following exceptions:

At the early prophase H, the two chromatids of each dyad looks like X, since they are conjoined by a common centromere. The four arms are widely separated. There is no rational coilling. The X-shaped dyads are quite longer than at telophase. The chromonemata are still not completely coiled.

The genetic constitution of the two chromatids of each dyad depends upon the kind and number of cross overs which took place in the phase I. If there is no crossing over, the dyads would consist of two identical sister chromatids of either paternal or maternal origin. At the end of the prophase II the nucleolus disappears. The nuclear membrane disappears and the acromatic figure is developed.

Metaphase II

As usual, all the chromosomes arrange themselves on the equator for a short duration. The centromeres touch the equator but their arms radiate out in different directions. Later on the centromere in each dyad divides into two sister centromeres.

Anaphase II

Sister centromeres separate to the poles, pulling with them the chromatids to which they are attached. The chromatids, however, become much thick and stout.

Telophase II

The chromosomes at each pole uncoil and thin out to form the nuclear net. Each group gets surrounded by a nuclear membrane. Nucleolus reappears. Thus two nuclei are recognised in each cell. This is soon followed by cytokinesis and two cells are formed from each haploid daughter cell. Thus as a result of meiosis four cells are produced, each with a haploid set of chromosomes, i.e. each contains just one members of each homologous pair.

Cytokinesis

Sometimes meiosis I is followed by cytokinesis and sometimes it is deferred until the end of meiosis II. The details of the process are same as during mitosis. Cell plate formation results into a tetrad of cells with reduced chromosome numbers.

The arrangement of cells in a tetrad is different in different organisms, but is characteristic of the species. Like mitosis, meiosis too is a dynamic process and its different stages merge into subsequent ones. There is no sharp demarcation between different stages. DNA synthesis takes place during interphase, prior to prophase I.

Duplication of chromosomes takes place during interphase itself and the chromosomes that appear during leptotene are double and not single. Number of chromosomes is reduced during meiosis I. If the reduction process is examined more closely, it becomes clear that there is qualitative reduction during meiosis I of those parts of chromosomes which are not involved in exchange. The homologous chromosomes or the maternal and paternal chromosomes segregate or undergo qualitative reduction during meiosis I. Their quantitative reduction takes place during meiosis II, when their centromeres divide and the sister chromatids separate. But the reverse is true of those parts of chromatids which are exchanged due to the formation of chiasmata.

For these parts quantitative reduction takes place during meiosis I and qualitative reduction during meiosis II. Thus, complete of both qualitative and quantitative reduction takes place as a result of meiosis I and II and none of them alone can bring about complete reduction. It is, therefore, not proper to say that meiosis I is reductional and meiosis II equational. Duration of meiosis varies a great deal in different organisms. So does the duration of different stages in an organism. Prophase I lasts for the longest duration.

Significance of Meiosis

For simplicity, this summary considers meiosis in a hypothetical cell containing only two pairs of chromosomes, one long pair and one short pair. At premeiotic G_1, these pairs are unassociated and randomly distributed in the nucleus. During premeiotic S, each chromosome replicates and becomes double at all points, and now contains two chromatids.

After G_2, as the condensation or leptotene stage begins, these chromosomes form into thick threads that become visible in the light microscope. Synapsis at the pairing or zygotene stage brings homologous chromosomes together, producing two tetrads in the nucleus, each containing two chromosomes and four chromatids.

During the recombination or pachytene stage, recombination occurs by the formation of exchanges or crossovers between the chromatids of the homologs. At the synthesis or diplotene stage, at least some decondensation, with RNA transcription, takes place. The chromosomes condense again at the final, recondensation or diakinesis stage of prophase I and enter into metaphase I. At metaphase I, the two chromatids of each homologous chromosome connect to microtubules leading to the same pole of the spindle.

Anaphase I therefore separates the two homologs of each pair and moves the haploid number of chromosomes to each pole of the spindle. Thus, only one long chromosome and one short chromosome are present at each pole, and pairs no longer exist. Note that at this point the chromosomes at the poles still contain two chromatids.

During interphase II and prophase II, two new spindles form at the metaphase I division poles. At metaphase II, the chromosomes move to the spindle midpoints and make microtubule attachments as in mitosis, in such a way that the two chromatids of each chromosome connect to microtubules leading to opposite spindle poles

Separation of the chromatids at anaphase II delivers the haploid number of chromatids to the poles. The four division products at telopbase II each now contain the haploid quantity of DNA and possess only two chromatids (now technically called chromosomes), one long and one short. If, in this hypothetical organism, meiosis and gamete formation produces eggs and sperm, the end products will each contain two chromosomes, one long and one short. Fusion of an egg and sperm nucleus in fertilisation rejoins the pairs and returns the chromosome complement to the G_1 level of premitotic cells.

Division of Sex Chromosomes in Meiosis

In many plants and animals, the males and females of a species may show differences in chromosome structure or number. The chromosomes that are regularly different in the males and females of a species are called *sex chromosomes*. In most of these species, cells in females contain a homologous pair of the sex chromosomes, and cells in males contain only one chromosome of this pair.

By convention, the sex chromosomes present in the female are called *X chromosomes*. In the male, the single representative of this pair, the X, may exist alone ("XO" males) or be associated with another sex chromosome, the Y, which is not found in females. In most animals and in the few plants with sex chromosomes, XX females and XY or XO males are the rule. In a few groups, such as the butterflies and moths, copepods, and birds, the situation is reversed, with XX males and XY females.

In the sex with two X chromosomes, the two X's pair normally and pass through the meiotic divisions in the same pattern as the rest of the chromosomes. In XY individuals, the X and Y chromosomes may or may not contain homologous segments capable of pairing during prophase I.

Whether or not pairing occurs, the X and Y line up on the spindle along with the rest of the chromosomes at metaphase I of meiosis. Both the X and Y contain two chromatids at this stage, as do the rest of the chromosomes. In most species, particularly where the X and Y undergo partial pairing and- recombination, the kinetochores of both chromatids of the X make connections to one pole, and the kinetochores of the Y make connections to the opposite pole, in a pattern similar to the remaining chromosomes. In this case, the X and Y separate and move to opposite poles of the spindle at anaphase 1. Anaphase II separates the chromatids of these chromosomes, so that the haploid products of meiosis receive either an X or Y chromatid. Less frequently, the two kinetochores of each sex chromosome make opposite spindle connections at metaphase I, along the lines of a mitotic division, with the result that both poles receive an X and a Y chromatid at anaphase I.

These make opposite spindle connections at anaphase II, usually after associating on the spindle, so that the final haploid products receive either an X or a Y. In individuals without a Y chromosome (XO individuals, common in insects) both chromatids of the single X usually go to the same pole of the spindle at anaphase I.

The other pole receives no X. At anaphase II, the nucleus receiving the X chromosome divides again and distributes one X chromatid to each of the two division products. The other nucleus, which receives no X at anaphase I, divides normally, yielding two nuclei with no X chromosome.

The reverse pattern of distribution is also noted, in which the two chromatids of the X chromosome make opposite spindle connections and separate at anaphase I. In fhe second meiotic division in this case, the single X chromatid moves to one of the two poles, leaving the other with no X. The result of the two patterns is the same Of the four nuclei resulting from meiosis, two will contain an X chromosome, and two will have no X (0 nuclei). At fertilisation, gametes containing an X or Y (or no Y) fuse to form the diploid zygote, restoring the XX or XY (or XO) complement of sex chromosomes.

Overall Patterns of DNA, RNA, and Protein Synthesis during Meiosis

The replication of a small fraction of the DNA, less than 1%, is delayed until meiotic prophase I, when it replicates during the pairing and recombination stages. Additional DNA synthesis specifically associated with the breakage and exchangeof segments between homologous chromatids is also detected during the recombination stage.

In some species, amplification of rDNA sequences, through their rapid synthesis and release from the chromosomes, also takes place during the early stages of prophase I. RNA transcription can be detected during prophase I in all organisms. This has been most clearly demonstrated in autoradiographs of cells exposed to ^{3}H-uridine, a radioactive precursor of RNA.

This transcription, which includes both bnRNA and rRNA, peaks in many species during the synthesis stage. This is particularly true in animal oocytes, where enormous quantities of ribosomes may be synthesized and packed into the cytoplasm in the synthesis stage. As cells reach metaphase I, incorporation of label drops to undetectable levels.

A brief period of RNA transcription, primarily of hnRNA, may occur again after telophase II, during development of the sperm cells in male animals. Incorporation of radioactive amino acids into proteins can also be detected throughout meiosis. Stern and his colleagues have shown that inhibition of protein synthesis by agents such as cycloheximide causes meiotic arrest and various abnormalities, depending on the stage at which cells are exposed to the inhibitor. For example, chiasma formation is blocked if cycloheximide is added to cells at the pairing or recombination stages. This blockage may result in part from an interruption of the synthesis of histone and nonhistone chromosomal proteins, which can be detected among ti e proteins assembled during prophase I.

Time and Place of Meiosis

There are three major variations in the time and place of meiosis in the life cycles of different eukaryotic organisms. The most familiar pattern takes place in animals, many protozoa, and a few lower plants. In this sequence, called *gametic* or *terminal* meiosis, the meiotic divisions occur immediately before gamete formation.

A second pattern, called *sporic* or *intermediate* meiosis, takes place in higher plants and in some lower plants and fungi. These organisms alternate in each generation between haploid and diploid individuals. Fertilisation produces the diploid *sporophyte* generation. At some point, meiosis occurs in the sporophyte, producing spores rather than gametes. Since the spores result from a meiotic division sequence, they are haploid and genetically diversified.

The spores on germination grow by mitotic divisions into functional plants that from the alternate haploid *gametophyte* generation. At maturity, the gametophyte generation produces eggs and sperm by differentiation of cells following an ordinary mitosis. All of the eggs or sperm from a single haploid gametophytic plant, since they arise through mitosis, are genetically identical.

Fusion of gametes returns the cycle to the diploid sporophyte generation. A third major variation in the time and place of meiosis is observed in some fungi and algae and in a few protozoa. In this form, termed *zygotic* or *initial* meiosis, the meiotic divisions take place immediately after fertilisation. Two haploid gamete nuclei designated as egg and sperm (or, since they are usually undifferentiated, simply as positive and negative) fuse to produce a diploid zygote. The zygote immediately enters meiosis, producing four haploid cells.

These eventually develop, after one or more mitotic divisions, into haploid spores, which germinate to give rise to haploid individuals. Eventually gametes are produced in these individuals as they are in the gametophytic generation of the higher plants, by differentiation of cells following an ordinary mitotic division. Individuals with initial meiosis have supplied the critical evidence establishing that recombination actually occurs at prophase I of meiosis.

Regulation of Meiotic Cell Division

The factors controlling the progress of cells through the meiotic cell cycle have proved as elusive as the controls of the mitotic cycle. In fungi and protists, essentially any cell can be induced to enter a meiotic rather than mitotic cell division by simple changes in the environment, usually toward less favourable growth conditions.

In higher plants and animals, induction of meiosis probably involves an interaction between hormones and the cell surface. Various internal responses, possibly including adjustments in the concentration of substances such as the cyclic nucleotides or Ca^{2+}, then trigger meiosis. The observations made to date suggest, as in mitotic cycles, that a variety of control mechanisms exist in different cells and species, ultimately acting on different segments of a complex collection of genes that are activated in sequence to trigger meiosis.

The few detailed studies carried out so far suggest that the first changes leading toward a meiotic cell division are imposed in cell lines cycling through mitosis several divisions before meiosis actually occurs. As we have seen, the duration of DNA replication gradually increases through several mitotic divisions preceding meiosis in *Lilium* and the mouse. And, in the lily, Stern and Hotta have observed that an increase in condensation of chromatin segments occurs in the nuclei of cell lines leading to meiosis as early as the G_2 period of the last mitotic cycle preceding meiosis.

Stern and Hotta's studies indicate that in the lily these early changes leading to a meiotic division can be reversed as late as the S stage preceding meiosis, so that a mitotic division occurs. They found that cells removed from their normal location in flower anthers and placed in a culture medium during premeiotic G_1 and S revert after a few days to an ordinary mitotic division. In this case, the cells replicate the small fraction of DNA left unreplicated in premeiotic S before dividing. Cells allowed to enter early G_2 before removal were unable to undergo a normal mitosis when placed in culture. Although spindles were formed, the chromosomes failed to separate at anaphase. Eventually these cells reverted to an interphaselike state without further replication or division.

Cells removed from the anthers later in G_2 entered an abnormal form of meiosis rather than mitosis in culture. These cells always exhibited irregularities in pairing during prophase, so that meiosis was never completed. Thus, the first changes irreversibly programming meiosis in these cell lines occur in G_2 of premeiotic interphase. These changes probably involve activation of a series of genes that triggers pairing and recombination.

These observations suggest that until G_2, the changes regulating entry into a meiotic division include primarily the usual genes controlling mitosis, with few genetic sites unique to meiosis active at these stages. Once into G_2, a series of genes is activated that fundamentally changes the mitotic sequence and produces the chromosome movements and activities of meiotic prophase I.

Since the second meiotic division proceeds much like a normal mitosis, the sequence at this stage is probably activated primarily by genes controlling mitosis, with few genetic sites unique to meiosis operating at this time.

The study of meiotic mutants supports the idea that most of the genetic sites controlling meiosis operate during prophase I and the first meiotic division, with fewer sites unique to meiosis active in premeiotic interphase or the second meiotic division. In *Drosophila,* for example, nearly 40 mutations affecting various steps in meiosis have been identified.

Of these, almost all cause defects in recombination and separation of homologous chromosomes during prophase I and the first meiotic division. Relatively few affect premeiotic DNA replication or the second meiotic division. For example, only 4 of the 40 known mutations interfere with steps in the second division in *Drosophila*. Other organisms show a similar pattern in the distribution of meiotic mutants.

All of these observations suggest that the evolutionary adaptations that established meiosis involved the development of pairing, recombination, and separation of homologous chromosomes during the first meiotic division. These meiotic adaptations are the most complex from the genetic standpoint and the most susceptible to disturbance through mutation.

Mutations that affect premeiotic replication and the second meiotic division, since most of the affected mechanisms of these stages are common to both meiosis and mitosis, are probably completely lethal because they cause drastic interference with all cell cycles of the individual. Thus, they are not likely to surface as mutations peculiar to meiosis.

Many parts of the meiotic sequence are not understood and are presently the subject of intensive research. Foremost of these are the processes of pairing and recombination during the early stages of prophase I and the patterns of DNA, RNA, and protein synthesis during late prophase I. These subjects are taken up in greater detail in the remaining sections of this chapter. The developmental changes leading to the formation of gametes following meiosis in animals and plants and the techniques for recombinant DNA research are briefly surveyed in Supplements 16-1 and 16-2.

MEIOTIC PAIRING

The Mechanism

The forces bringing homologous chromosomes together into tightly synapsed pairs during the pairing or zygotene stage are only very incompletely understood. Detailed anal) sis of the pairing mechanism reveals that it takes place in several stages. First, the homologs move from their scattered locations and approach until they are separated by a distance amounting only to several hundred nanometers.

At this point, according to an analysis of pairing in *Locusta* by P. Moens (1969b), the first physical connections can be seen between the pairing chromosomes. These connections take the form of fibers about 300 nm long that extend between the homologs. As these connections develop, the chromosomes move closer together, possibly through contraction or sliding of the connecting fibers, until they are separated by a fairly regular distance of about 200 nm, or 0.2 μm.

Approach to this distance is accompanied by the assembly of the synaptonemal complex between the homologs. Complete assembly of the synaptonemal complex at all points establishes the second stage of pairing. At this stage, pairing is still relatively gross, at least in the molecular sense, and limited to alignment of homologous chromosomes in chromomere-by-chromomere register.

In the third and final stage, close molecular pairing of DNA molecules originating from the two homologs takes place, presumably by the movement of DNA from opposite chromomeres into the central region of the synaptonemal complex. At this final level of pairing, homologous DNA molecules must approach to within distances close enough to allow base pairs to form, on the order of a few tenths of a nanometer.

The forces accomplishing the first stage of pairing are unknown. Initial approach of the homologs, as proposed by D. E. Comings and others, may involve the nuclear envelope, since the first close alignment of homologs is frequently observed where they connect to the inner membrane of the nuclear envelope.

Movement of the chromosomes by the envelope is not without parallels in other cellular systems. In some primitive organisms, the chromosomes are evidently separated in mitosis entirely by movements of their points of attachment to the nuclear envelope. Presumably, this movement is accomplished by active sliding or conformational changes in proteins embedded within the membranes of the nuclear envelope or associated with the nuclear envelope surface.

Several observations make it clear, however, that the initial approach of the homologs in pairing cannot be due entirely to the activity of the nuclear envelope. Initial approach and pairing also take place at scattered sites within the nucleus, at some distance from the nuclear envelope. Other observations come from the study of triploids and of cells in which a chromosome segment has been translocated from one set of homologs to another.

Triploids have three copies of each chromosome, instead of two as in diploids. During meiotic prophase, the three homologs associate in such a way that, at any given point, only two of the three chromosomes come together and pair. Which two pair at any point is random, and, most significantly, alternates frequently between any two of the three homologs along the chromosomes. These switches between pairing partners would not be expected if pairing begins and extends only from the nuclear envelope.

The pairing configurations noted between translocated chromosomes lead to the same conclusions. In the translocations of interest, a chromosome segment is broken and joined to the end or middle of another, nonhomologous chromosome of the set. Such translocations can be readily induced by exposing cells to X rays, which cause random chromosome breaks that subsequently rejoin in various patterns.

If the translocation is long enough to make switching between the chromosomes mechanically feasible, the translocated segment will still pair with the corresponding homologous segment of its original pairing partner. Once again, if all pairing begins and extends only from the nuclear envelope connections, such changes between pairing partners in the interior of the nucleus would not be possible.

Another hypothesis for the mechanism of initial pairing maintains that pairing is actually set up during premeiotic interphase, before the chromosomes begin to condense into visible threads at the beginning of prophase I. While the findings are somewhat controversial, apparent premeiotic pairing of this type has been reported in a few plant species, usually on the basis of associations between the patches of condensed chromatin that occur in quantity in nuclei during premeiotic interphase.

Nothing definite can be established from these observations, since the scattered blocks of condensed chromatin in the species observed cannot be traced and identified as parts of specific homologous or nonhomologous chromosomes. Where the blocks can actually be identified with specific homologs, in fact, indications are that there are no pairing associations between homologous chromosomes before meiosis begins.

For example, in the insect *Stauroderus,* all of the homologous pairs have conspicuous, readily identified segments that condense and become visible during premeiotic interpbase. Tracing the locations of these segments shows that they are randomly distributed, with no tendency to approach and pair, during the interphase before meiosis. The pattern of meiosis in organisms with the initial pathway also makes premeiotic pairing seem unlikely as a prerequisite for synapsis during prophase I.

In these species, the homologous chromosomes are in separate nuclei during premeiotic interphase. At best, the possibility of premeiotic pairing as the basis for synapsis in prophase I merely begs the question, since the longdistance forces that might act to bring the widely scattered homologs together are as difficult to explain at interpbase as at prophase I. In fact, no known molecular forces can act over the distances separating the homologs before pairing, which may amount to many micrometers.

The forces bringing the chromosomes into the second stage of pairing, in which the synaptonemal complex assembles and the homologs are brought into chromomere-by-chromomere register, have proved equally elusive and difficult to explain.

Whether the synaptonemal complex contributes to the forces bringing the homologs together or functions simply as a stabilising structure once pairing has been accomplished is unknown. Stern and Hotta's research into the biochemistry of meiosis has shown that the DNA synthesized during the pairing stage is necessary for the mechanisms bringing the bomologs to the second stage of pairing. Stern and Hotta cell this DNA replication Z-DNA synthesis, because it occurs during the zygotene or pairing stage. In *Lilfum*, the species studied by Stern and Hotta, Z-DNA synthesis amounts to about 0.3% of the total DNA.

Inhibition of this synthesis inhibits pairing and interrupts formation of the synaptonemal complex. These investigators have proposed the following model for synapsis during the pairing stage, based on their observations of Z-DNA synthesis. According to their interpretation, the small amount of DNA left unreplicated during premeiotic S represents short, single-stranded segments that are distributed throughout the chromosomes.

Some of these are trapped randomly in the first elements of the synaptonemal complex attaching to the chromosomes. These first elements, the lateral or side elements of the complex, attach during the condensation stage before pairing begins. As the complete synaptonemal complex forms, base pairing between the single-stranded segments trapped within the complex brings the chromosomes of opposite homologs into register. Other investigators have proposed that RNA or protein molecules capable of recognising specific DNA sequences are built into the synaptonemal complex. According to this idea, these molecules are instrumental in recognising segments of opposite, homologous chromomeres and bringing them into alignment during pairing.

The forces bringing the chromosomes into the third stage of synapsis, in which DNA molecules enter into close molecular pairing, form a part of the mechanism of breakage and exchange of chromosome segments during recombination. Before considering the evidence and models for this stage of pairing, we must take up some details of the morphology and assembly of the synaptonemal complex, a structure that is probably intimately involved in the process of close molecular pairing and recombination.

The Synaptonemal Complex and Chromosome Pairing

Structure of the Synaptonemal Complex

The synaptonemal complex (also called the *axial core*) appears between the synapsed homologs in all eukaryotes in which pairing and recombination take place. The basic morphology of the complex in sections made parallel to the long axis of the chromosomes is remarkably uniform in different species and consists of three parallel, dense elements separated by less dense spaces, all of regular dimensions.

The two *lateral elements* at either side of the complex, which vary in width in different species from 20 to 60 nm, appear to be made up of dense granules or fibers of slightly larger diameter than the surrounding chromatin fibers. Frequently the lateral elements can be seen to be split longitudinally into two halves. The chromatin surrounding the lateral elements may be either aggregated into dense, irregular masses or so diffusely extended that no distinct outline is discernible. In most organisms, the lateral elements contain no regularly ordered structures.

However, in the ascomycete fungi and a few grasshoppers, the lateral elements show a pattern of regular crossbands. The lateral elements are separated by a regular central space that varies in different organisms over a narrow range from about 90 to 120 nm. Along the middle of the central space lies the *central element,* separated by an equal distance from either lateral element.

The central element takes two basic forms. In most plants and animals, the element is indistinct and apparently structureless. In some insects and in the ascomycete fungi, the central element takes on a more ordered appearance, appearing ladderlike, with two or three longitudinal, parallel side components connected by regularly spaced cross-striations. Where the complex twists so that the plane of section passes vertically through the ladderlike array, the central element appears as five to six layers of dots in longitudinal rows.

The pattern of these dots, which are probably cross sections of the runglike cross-striations, suggests that five to six ladders stack vertically to form the central element. In all forms of the synaptonemal complex, very thin fibers cross the central space and connect the central element with the lateral elements.

These *transverse fibers,* which range in diameter from 6-7 nm down to invisibility, lie more or less at right angles to the lateral and central elements. Cross sections show that the lateral elements are spaced on either side of the central element, giving the impression that the synaptonemal complex is a flat, ribbonlike structure. The side elements of the complex thicken where the chromosomes attach to the nuclear envelope, forming dense, caplike structures closely fused to the nuclear envelope membranes.

Molecular Constituents of the Synaptonemal Complex

Much interest centers on the molecular structure of the synaptonemal complex, since it is in the right place at the right time to be directly involved in pairing and recombination. This problem has been approached by a variety of analytical techniques for electron microscopy, none of which can be depended upon for completely definitive results.

The clearest indications of molecular structure of the complex have been obtained from the Bernhard EDTA staining technique, which stains RNA-containing structures selectively. The lateral elements are intensely stained by this technique, and the central element to a lesser degree, indicating that RNA is present in these regions of the synaptonemal complex.

Sensitivity of the lateral and central elements to RNAse digestion, noted in many studies, also suggests that both structures contain RNA. The entire complex, including lateral and central elements and the transverse fibers, is also readily digested by proteases such as trypsin and pronase. Taken together, these observations indicate that the synaptonem; complex is a ribonucleoprotein framework that extends betwee the synapsed homologs. Several techniques suggest that at least some of the material of the lateral elements is a basic protein, possibly of the histone type.

Phosphotungstic acid, a stain with some affinity for proteins containing basic amino acids, stains the lateral elements densely. Another technique, involving exposure of the synaptonemal complex to silver ions at pH 8-9, also stains the lateral elements darkly. According to Westergaard and von Wettstein (1972), the technique stains basic proteins with a high arginine content because arginine residues arc expected to react most strongly ,i ith the silver ions at this pH.

The fact that the appearance and disappearance of a specific meiotic histone in *Cilium* is closely coordinated with the assembly and disassembly of the synaptonemal complex during prophase I also suggests that the basic protein of the lateral elements may be a histone.

The presence of DNA in the synaptonemal complex has proved more difficult to establish. On staining paired chromosomes with indium, a heavy metal ion considered to have some selectivity for DNA, J.R. Coleman and M.J. Moses noted that the chromatin fibers surrounding the complex, as well as the inner border of the lateral elements facing the central element, became more dense.

Very faint density could also be seen in the thin transverse fibers connecting the lateral and central elements. Digestion of the cells with DNAse before exposure of the chromosomes to indium eliminated all staining in the chromatin, lateral elements, and transverse fibers, indicating that the indium staining was actually due to the presence of DNA.

Staining the DNAse digested chromosomes with lead ions, a treatment considered to act as a general stain for proteins, added density to all of the elements of the synaptonemal complex and revealed that the basic framework was still intact after exposure to the DNAse enzyme. Resistance of the basic elements of the synaptonemal co:mplex

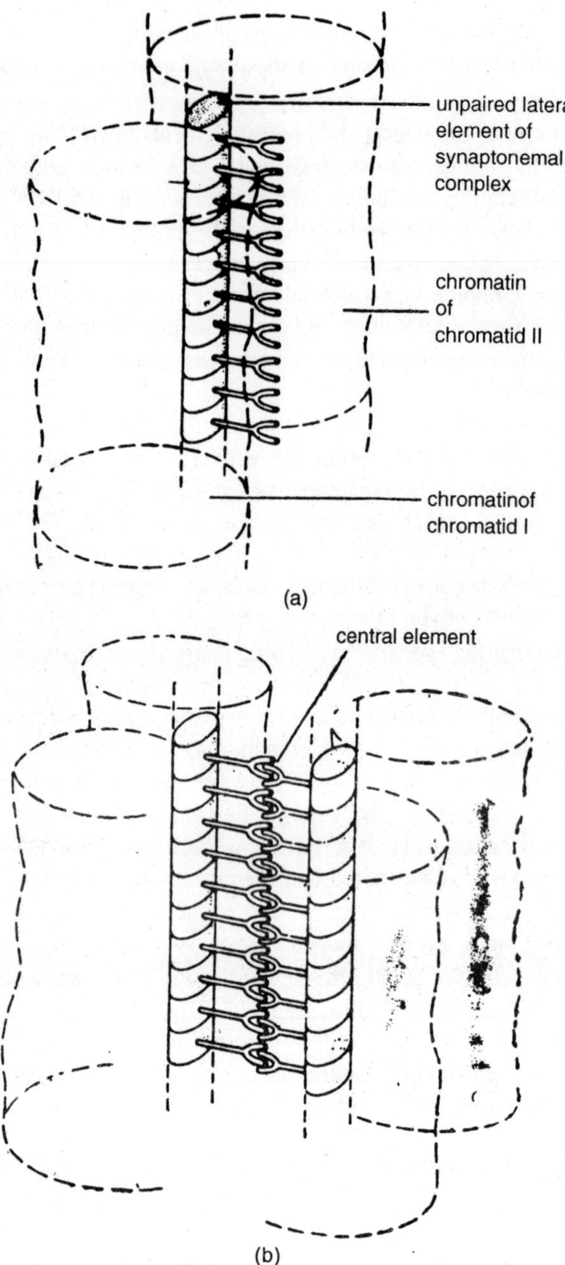

Figure 3.5: Formation of the synaptonemal complex as the homologs pair.

to DNAse digestion has also been generally observed in other studies. Taken together, these results suggest that the framework of the synaptonemal complex, including the lateral and central elements and the transverse fibers, consists of protein or an RNA-protein complex. Any DNA present in the structure probably represents chromatin originating from the homologs, threading through the lateral elements and extending along the transverse fibers into the central element.

Formation and Disassembly of the Synaptonemal Complex

Single, unpaired lateral elements appear as the first visible parts of the synaptonemal complex during the condensation stage of prophase 1. Careful analysis of cross sections of the unpaired homologs by Westergaard and von Wettstein shows that at this stage, the single lateral element is located in the split between the two chromatids.

According to Westergaard and von Wettstein, location of the unpaired lateral element in this position and the close association of the chromatin fibers of the chroratids with this element account for the fact that the separation between sister chromatids is invisible during the early stages of prophase I. Interestingly, staining the lateral elements by the Bernhard EDTA technique produces negative results as long as the lateral elements are in the unpaired state.

This suggests that the RNA associated with the lateral elements does not appear until the chromosomes are fully paired and the complex is completely formed, and makes it seem unlikely that RNA recognition molecules could be active in the initial pairing mechanism bringing the lateral elements together. During the pairing stage, the lateral elements of homologous chromosomes move together until they are separated by approximately 100 am.

As they come together, the central element appears. The pattern of its formation suggests that the central element assembles through the association of short, fibrous projections that extend at right angles from the lateral elements, much like the teeth linking together as a zipper closes. The synaptonemal complex persists between the homologous chromosomes as long as they remain closely paired.

The separation of the homologs at the close of the recombination stage is correlated with the disappearance of the complex, which remains only at points where the homologs are held together by chiasmata. Disassembly of the complex follows one of several patterns, depending on the species. In some, disassembly proceeds in the reverse of complex formation during pairing. The central element disappears first, as the chromosomes separate, and single, unpaired lateral elements persist briefly along the unpaired homologs. In other species, the complex simply disintegrates or is stripped from between the homologs while still more or less intact.

In a few organisms, the elements of the synaptonemal corrplex reassemble into aggregations called *polycomplexes* after disassembly from the chromosomes. These structures, which appear in insects and a few other species, contain masses of alternating lateral and central elements. These repeating elements may number from 3 or 4 to as many as 60. The collections of elements may be relatively flat, with no greater thickness than an ordinary synaptonemal complex, or extend in all directions, so that the mass takes on the appearance of a mulfilayered sandwich.

Even more bizarre configurations occur, including tubular complexes semicrystalline arrays. Usually, there are no connections between polycomplexes and chromatin fibers. In most organisms, polycomplexes disappear by the end of prophase I. However, in some instances they can still be detected in males as late as the early stages of sperm formation, after all of the stages of meiosis have run their course. Since polycomplexes do not appear until close pairing ceases, it is generally agreed that they are assembled fron elements of the synaptonemal complexes removed from the chromosomes.

This conclusion and a series of additiona observations suggest that the synaptonemal complex is forme(from subunits that have the capacity to self-assemble. On line of supporting evidence comes from the appearance of polycomplexes in the nuclei of nonmeiotic "*nurse*" cells *it Aedes* mosquito ovaries. These cells, which surround the developing oocytes, evidently synthesize and assemble subunit,, of the synaptonemal complex in response to the same metabolic triggers that influence complex assembly in the nearby meiotic cells.

Another line of supporting evidence comes from an observation of early polycomplex formation in *Ascaris.* In this organism, polycomplexes appear in the cytoplasm during the condensation stage, before pairing begins. Late in this stage the polycomplex mass attaches to the outside of the nuclear envelope. As the chromosomes pair and the synaptonemal complex forms between the homologs, the mass of polycomplexes disappears from the cytoplasm.

As Bogdanov suggests, this pattern of appearance and disappearance of the polycomplexes may result from the self-assembly of subunits of the synaptonemal complex, which are synthesized in quantity in the cytoplasm before pairing begins. This capacity for self-assembly may also account for the appearance of the synaptonemal complex between nonhomologous chromosome segments under unusual circumstances, as in associations between chromosomes noted in haploids.

In some haploids, as in -haploid tomatoes, apparently normal synaptonemal complexes form in the reproductive cells of the ovary or anther, even though homologous pairs are not present. The complexes form instead between segments of the chromosomes that fold back on themselves, or between segments of separate chromosomes that make fortuitous contact.

In this instance, subunits of the synaptonemal complex are probably synthesized, and reach high concentrations in the haploid cells,. under the influence of the metabolic signals triggering meiosis in normal tissues at this time. Fortuitous contact of chromosome segments promotes self-assembly of the subunits into synaptonemal complexes in these regions, even though the chromosome segments are nonhomologous.

Observations of this type, incidentally, also indicate that the initial assembly of the synaptonemal complex is a nonspecific process that does not depend on the presence of homologous DNA or RNA recognition sequences. The observations in this section suggest that assembly and' disassembly of the synaptonemal complex proceeds according: to the pattern. The subunits of the synaptonemal complex are considered to be monomers that can polymerize to form the elements of the complex.

The reversible conversion between the subunit pool and assembled synaptonemal complexes exists as a dynamic equilibrium that can be adjusted in either direction by changes in nuclear conditions, much like the reaction polymerising spindle microtubules. At premeiotic G_2 or early prophase I, subunits of the complex are synthesized in the cytoplasm in response to metabolic signals triggering meiosis. These cross the nuclear envelope and reach high concentrations in the nucleus.

During the condensation stage, factors associated with the chromosomes promote the assembly of the subunits. into lateral elements in the split between the homologs. Pairing, possibly induced by the tendency of the subunits to selfassemble, brings the lateral elements together to form the central element and complete the synaptonemal complex. The initial pairing configuration achieved may simply bring opposite chromomeres into rough alignment.

Single-stranded Z-DNA sequences or RNA or protein sequence recognition molecules may then be integrated into the complex to promote close alignment of the homologs. Once close alignment is achieved to complete the second stage of pairing, the synaptonemal complex provides an organising framework for the enzymes and factors involved in molecular pairing and recombination.

After recombination, a change in nuclear conditions pushes the equilibrium in the direction of the monomer pool, and the synaptonemal complex disassembles. In most organisms, the subunits disperse at this time into the nucleoplasm or cytoplasm and are degraded. In a few species, the subunits reassemble locally into polycomplexes and are stored temporarily in this form until breakdown commences.

Role of the Synaptonemal Complex

Although significant progress has been made in unraveling some of the molecular events associated with genetic recombination, such studies have not revealed much about the relationship between the structural organisation of meiotic chromosomes and the molecular events of recombination.

Little is known, for example, about chromatid organisation during the recombination process or the mechanism responsible for the close pairing of the homologous chromosomes. One structure thought to be intimately involved in these events is the *synaptonemal complex. This* complex, which occupies the space between paired homologous chromosomes, was discovered independently in 1956 by Montrose Moses and Don Fawcett.

In electron micrographs the synaptonemal complex can be seen to contain an electron-dense *central element* that runs lengthwise between the two chromosomes, plus two parallel *lateral elements,* one associated with each of the two homologs. The central and lateral elements are connected to each other by a series of *transverse fibers.* Synaptonemal complexes have been detected in almost all eukaryctic cells undergoing meiotic division, although variations in size and organisation do occur.

The overall width of the complex, for example, is about 160 nm in tomatoes, 180 nm in mammals, and 200 nm in crickets. The central element generally ranges from 12-20 nm in diameter, and is separated from the lateral elements by spaces of approximately 40 nm. Cross sections through the central element reveal the presence of fibers about 5-7 5 nm in diameter, while the lateral element consist of granules and fibers that are slightly larger.

In some cell types the termini of the synaptonemal complex appear to about upon the inner membrane of the nuclear envelope. The relationships among these various elements of a "typical" -synaptonemal complex are summarised.

Cytochemical studies have revealed that the bulk of the synaptonemal complex is accounted for the protein, though small amounts of DNA can be detected in the lateral elements. This DNA is thought to be derived from the chromatids with which the lateral elements are closely associated. The transverse fibers crossing the space between the central and lateral elements may also contain some DNA.

This overall arrangement suggests that the synaptonemal complex is composed of a framework of protein within which DNA-containing chromosomes fibers is various stags. of uacoiling are interspersed. The first step in the development of the synaptonemal complex occurs during the leptotene stage of prophase I, at which time a lateral element is formed adjacent to each homologous chromosome.

At zygotene the two chromosome of each homologous pair come together, leaving a space of about 100 nm between their respective lateral elements. The central element then forms in this space, generating a fully developed synaptonemal complex. This structure is retained until the beginning of diplotene, at which point is disassembles as the homologous chromosomes separate.

In some cells remnants of the synaptonemal complexes persist through the second meiotic, division, forming aggregates known as *Polycomplexes.* A number of indirect arguments suggest that the synap nemal complex plays an important role in genetic recombination. First of all, the synaptonemal complex is formed the same time that recombination takes place, and its physical location between the opposed homologous chromosomes con ponds to the region where crossing over occurs.

More direct support for the notion that recombination involves the synaptonemal complex has emerged from experiments in which genetic recombination was prevented by exposing cells to high temperatures or drugs that block DNA synthesis. Under such conditions synaptonemal complexes are not formed. Likewise synaptonemal complexes have been found to be absent from the meiotic cells of male fruit flies, which are unusual in that they do not carry out genetic recombination. Although a correlation between the presence of a synaptonemal complex and the occurrence of genetic recombination clearly exists, the exact role played by this structure is not known.

Since synaptonemal complexes occasionally form between nonhomologous chromosomes, it is thought that they cannot be responsible for precise alignment of the chromosomes. An alternative possibility is that the synaptonemal complex mediates a relatively crude alignment between

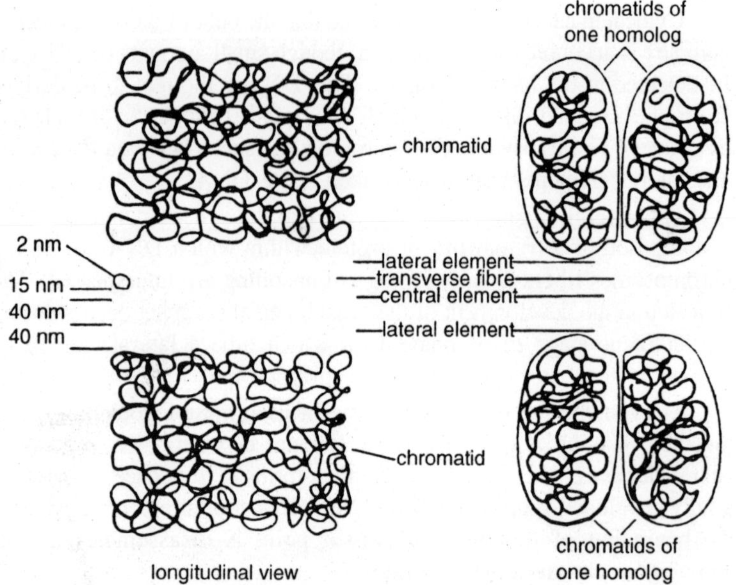

Figure 3.6 : Schematic model of the synaptonemal

two homologous chromosomes, bringing comparable regions of their respective DNA molecules into the general vicinity of one another. It has been suggested that each chromatid then undergoes localised uncoiling, generating a series of looped out DNA regions that project into the area of the central element.

The transverse fibers observed in the space between the central and lateral elements may represent such loops. The molecular events associated with DNA strand exchange would then take place between these looped-out regions. Such exchange can only occur, of course, when the looped-out regions of DNA contain homologous base sequences, since base pairing between complementary DNA sequences is a crucial ingredient in the molecular models of DNA strand exchange discussed in the previous section.

If the simultaneous looping out of comparable regions of opposing chromatids is a relatively rare event, this might explain why recombination occurs so infrequently. In spite of the evidence implicating the synaptonemal complex in genetic recombination, some observations are difficult to reconcile with this general notion. It has been reported, for example, that certain strains of *Drosophila* carry out genetic recombination in the absence of a synaptonemal complex.

Although this may represent a special case in which the close chromosomal pairing required for recombination is accomplished by another mechanism, an alternative possibility is that the synaptonemal complex normally performs a subsidiary, nonessential role in genetic recombination. It might, for example, be responsible for imparting a structural organisation that ensures that genetic exchange occurs between chromatids of opposing homologs, but not between chromatids of the same homolog.

THE MECHANISM OF RECOMBINATION

Recombination in Eukaryotes

As a consequence of recombination, some of the chromosomes at the end of meiosis differ in sequence from their counterparts in somatic cells of the same species. The changes, which arise through physical exchanges of segments between homologous chromosomes. Consider a homologous pair in which both chromosomes contain the genes A, *a,* and B, *b* at different sites on one chromosome arm. One member of the pair has the alleles A and B at these locations.

As a results of replication of premeiotic interphase, both chromatids of the chromosomes will have the A and B alleles at these sites. The opposite member of the homologous pair has the alleles a and *b* at the corresponding locations. At this stage, these chromosomes are identical in sequence to chromosomes of the same homologous pair in any cell of the same organism. During prophase I of meiosis, the homologs pair closely. If recombination occurs between these genes, two of the four chromatids at the close of meiosis will contain the new combinations *A-b* and a-B. The process of recombination in eukaryotes has several characteristics that can be demonstrated by genetic crosses or biochemical tests.

Recombination results from physical exchanges between homologs, which, as will be demonstrated, must involve a process of breakage and reunion of chromosome segments. Each exchange event, involves only two of the four chromatids of a tetrad. As a result, the, products of recombination consist of two chromatids of unchanged sequence (the *parentals)* and two with altered sequence (the *recombinants).*

The exchanges between chromatids become visible as chiasmata or crossovers at later stages of prophase I. In most species, the positions at which an exchange may take place are randomly located on any chromosome arm.

Exchanges are reduced or absent in some regions, however, such as the primary constrictions or in concentrations of heterochromatin containing highly repetitive sequences. The occurrence of one exchange reduces the probability that another may occur close by on the same chromosome arm. This effect is called *positive interference*. Normally, the chromosomes exchange segments in recombination with perfect precision, so that neither of the recombinant chromatids winds up with a single nucleotide too many or too few.

Establishing the Breakage and Exchange Mechanism of Recombination

The characteristics of recombination described up to this point were identified primarily through the experiments of classical genetics. However, conclusive evidence that recombination occurs by breakage and exchange of chromatid segments during prophase I was not developed until the 1960s. The definitive experiment was carried out by J.H. Taylor (1965), using meiotic cells of the grasshopper *Romalea*.

Taylor injected grasshoppers with a pulse of radioactive DNA precursor (^{3}H-thymidine) and followed the distribution of label in the two meiotic divisions by autoradiography. The cells of interest in the experiment are those taking up label in the interphase just before the last premeiotic mitosis. Before replication begins, none of the DNA in these cells is labeled. DNA synthesis takes place in the presence of the injected label.

As a result of semiconservative replication, these cells enter mitotic metaphase with one nucleotide chain of each DNA molecule labeled and one unlabeled. As a consequence, all chromatids will show the presence of radioactive label. After this last premeiotic division is complete, these cells enter the interphase before meiosis. At this point, the labeled precursor has been removed and replication takes place in

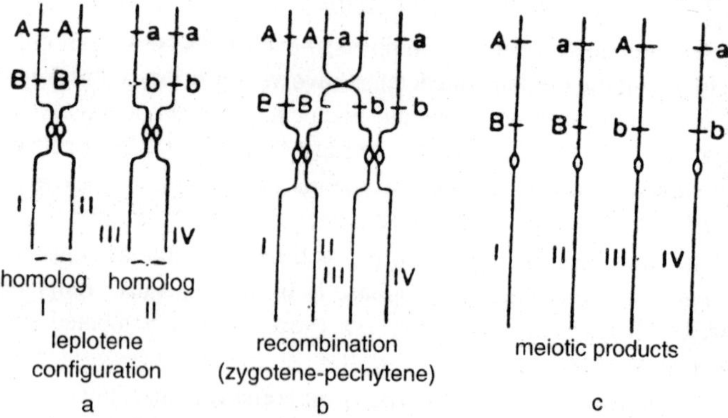

Figure 3.7 : The effect of recombination in mixing the chromatids entering meiosis into new combinations.

unlabeled medium.

Semiconservative replication of each chromosome will produce two chromatids, one showing the presence of label, and one completely unlabeled. Up to this point, therefore, the results are the same as those obtained in Taylor's demonstration of semiconservative replication by the same techniques. If no physical exchanges occur between chromatids as a part of recombination during meiotic prophase I, any single chromatid at metaphase of the subsequent meiotic division should be either completely labeled or unlabeled.

If recombination takes place by physical cha[ex]changes of chromatid segments, chromatids that contain labeled and unlabeled segments at the subsequent meiotic divisions should be detectable. Chromatids with segmented label were actually observed by Taylor in ratios approaching the observed recombination frequency. Because these could arise only by breakage and exchange of segments between labeled and unlabeled chromatids, Taylor's experiment established that breakage and exchange occur during meiosis in higher organisms.

Time and Place of Recombination

The final link in the evidence for the breakage and exchange hypothesis concerns the time at which recombination occurs. Considerable argument has hinged on this facet of recombination, In the early 1950s, J.H. Taylor and H. Swift independently demonstrated that the bulk of DNA replication takes place during premeiotic interphase and not during prophase I of meiosis, as some investigators had maintained.

Is recombination associated with DNA replication during premeiotic interphase, or does it take place after replication, during prophase I of meiosis, as predicted by the breakage-and-exchange model? The definitive experiment establishing that recombination occurs during prophase I came from the research of J.M. Rosen and M. Westergaard, who worked with *Neottiella,* a fungus in which zygotic or initial meiosis occurs.

In this fungus, an ascomycete, nuclei are haploid in the somatic tissues. The haploid gamete nuclei, derived from somatic nuclei after an ordinary mitosis, fuse to form the zygote. Meiosis then takes place immediately. Rossen and Westergaard found by measuring the quantity of DNA per nucleus that replication occurs while the nuclei are still haploid, before they fuse to form the zygote. Because the nuclei are still haploid, no recombination can occur at this time.

After fusion of the gametes, meiosis occurs immediately, with typical synapsis of homologous pairs during meiotic prophase I. The

Table 3.1: Showing Differences between Mitosis and Meiosis.

	Mitosis		*Meiosis*
1.	Prolonged resting stage.	1.	This stage is very short.
2.	The prophase is short without any substages.	2.	The prophase is subdivided into six stages, viz,. Proleptotene Leptonema, Zygonema, Pachynema, Diplonema and Diakinesis.
3.	The chromosomes split longitudinally into sister chromatids during the preceding resting stage. From early prophase they appear as double tetrads.	3.	The chromosomes do not split longitudinally.
4.	The split homologues are not attracted towards each other.	4.	At zygonema or succeeding stages, in prophase of first division, the homologous chromosomes are attracted to-wards each other and they form bivalents.
5.	Synapsis and tetrad stages do not occur.	5.	The chromosomes split after synapsis to form sister chromatids which are arranged in tetrad,
6.	No chromosomal crossing over takes place.	6.	The crossing over is a prominent feature which results in exchange of genic material.
7.	The chromatids are coiled in plectonemic manner.	7.	Coiling is paranemic type, and much evident.
8.	Doubling of chromosomes takes place early in pro-phase.	8.	Doubling of chromosores occurs in pachytene stage of prophase.
9.	Nucleolus and nuclear membrane disappear later on and the formation of nuclear spindle is completed with the commencement of metaphase.	9.	The nuclear membrane and nucleolus disappear and the formation of acromatic figure is completed with diakinesis simultaneously.
10.	The nucleus enlarges to some extent.	10.	The nucleus is greatly enlarged.

11. At metaphase, the centrosome of each chromosome divides and the sister chromatids move towards the oppo-site poles. The somatic number of chromosomes is maintained without any reduction.	11. At metaphase I, the centres-ome does not divide but the homologous chromosomes move towards opposite poles. The chromosome number is reduced to half by reduction division.
12. The chromosomes are seen in dyad consisting of two chrom-atids.	12. The chromosomes are in the form of tetrads having four chromatids.
13. Each chromosome of pater-nal and maternal origin is longitudinally divided and the separation to opposite side results in identical cells at the end of mitosis,	13. At metaphase I, the orienta-tion of chromosomes at the equatorial plate is in such pattern that either paternal or maternal chromosome goes to either pole. The orientation of one pair is independent of the other.
14. During orientation of dyads the centromeres face to-wards the equator and the arms directed towards the poles.	14. During the orientation of tetrads, the centromeres are placed equidistant and face the pole while arms are directed towards the equator.
15. The centromere of each bival-ent undergoes division at metaphase.	15. No division of centromere occurs, but homologous chromosomes simply separate to form two dyads.
16. The initial movement of chromosomes at late metaph-ase is due to repulsion between the centromeric parts.	16. The main repulsive force lies between the centromeres of homologous chromosomes.
17. At anaphase the chromosomes are less thick and short.	17. The chromosomes , are very thick and short.
18. At anaphase further separation of chromosomes is due to the formation of stem body.	18. Most of the changes are similar to mitosis.
19. At telophase two daughter nuceli, each with diplid num-ber of chromosomes are recon-stituted.	19. At this phase two daughter nuclei are haploid.

20. There is no evidence of second division in mitosis.	20. There are two division for each meiotic division. The second division follows after a short gap, interkinesis. The second division is homotypic by nature, and results into four cells, in the form of tetrads.
21. In the mitosis neither the chromosomal number nor the quality is changed and the daughter cells carry 2a chromosomes.	21. At the end of meiosis each of tetrads is haploid and quality is changed.
22. This division takes place in growing somatic cells.	22. This division takes place only in gonodial cells and result in the formation of gametes.

presence of chiasmata at later stages of prophase I was taken as evidence that recombination takes place in this species. Since the exchanges responsible for these chiasmata could occur only during prophase I of meiosis, the only time both chromosomes of each homologous pair are present together in the same nucleus, the experiment effectively eliminated the possibility that recombination occurs in premeiotic S during replication of the DNA.

Equivalent experiments have also been carried out by others in the alga *Spirogyra* and in the mushroom *Coprinus,* which also follow the initial pathway for meiosis and replicate their DNA while still in the haploid state. One line of argument once used to support recombination during premeiotic interphase and still used to some extent cited the fact that a variety of agents changing the frequency of recombinants have their most pronounced effects if applied during premeiotic S.

Exposing cells in premeiotic S briefly to high temperatures or to inhibitors of DNA synthesis, example, significantly alters the frequency of recombination. Rather than proving that recombination takes place during premeiotic interphase, these observations may simply indicate that the normal progress of the specialised pattern of DNA synthesis noted in premeiotic S, in which a fraction of the DNA is left unreplicated, is essential for recombination later in meiotic prophase I.

The experiments outlined in this section thus establish conclusively that recombination (1) occurs by breakage and reunion of chromatids, and (2) takes place at prophase of meiosis after replication of the chromosomes.

TYPES OF MEIOSIS

There are three types of meiosis, *zygotic or initial meiosis, gametic or terminal meiosis and sporic or indeterminate meiosis.*

(a) Zygotic or initial meiosis (haplotone pattern) occurs in lower plants. Fertilisation is immediately followed by meiosis, giving rise to cells with the haploid chromosome number. The zygote is the only diploid stage in the life cycle.

(b) Gametic or terminal meiosis (diplotonic pattern) occurs in animals. The cells resulting from the two meiotic divisions are transformed into sperms or eggs without further cell division. The process is as follows. Primordial germ cells divide repeatedly to give rise to many generations of daughter cells, spermatogonia (in males) or oogonia (in females).

The spermatogonia and oogonia are transformed into primary spermatocytes and primary oocytes, respectively.. Each primary spermatocyte undergoes the first maturation (metiotic) division and froms two secondary spermatocytes. Each secondary spermatocyte undergoes the second maturation (meiotic) division to give rise to two spermatides, which are then transformed into spermatozoa.

In the female during the first meiotic division the primary oocyte, into a large secondary oocyte and a smaller cell, the first polar body. The secondary oocyte undergoes the second meiotic division and divides into an ootid and a second polar body. The first polar body may or may not divide into two. The ootid is the ovum proper.

(c) Sporic or indeterminate meiosis (diplohaplotonic pattern) is characteristic of higher plants and some thallophytes, but is not known in animals. In the diploid sporophyte, meiosis takes place at some indeterminate point between the zygote and gamete formation.

As a result of meiosis haploid megaspores and microspores are formed. The megaspores and microspores undergo repeated mitotic divisions and form the male and female gametophyte, respectively. Certain cells of the gametophyte differentiate and from the gametes. In plants the cells that undergo meiosis are the microspore mother cell and the mesgaspore mother cell.

ROLE OF CENTRIOLE DURING MEIOSIS

The centrioles perform precisely controlled synchronous function during meiosis. Where meiosis occurs in two steps, that **is,** two nuclear divisions take place in sequence, the centriole reproduces and functions to from two achromatic figures.

In the first division of meiosis, the centromeres do not duplicate but rather homologous chromosomes pair together and move to opposite poles, while in the secor d meiotic division the centromeres of chromatids are separated. A scheme for relating various key events is presented briefly as under:

(*a*) Nonduplication of centromeres in meiosis-I,

(*b*) Nonduplication of chromosomes in meiosis-II, and

(*c*) Duplication and function of centrioles in both divisions.

By a complete repression of the duplication, simultaneous function of the achromatic apparatus brings about the reduction. The cetriolar function is super-imposed upon the duplication process of the chromatic apparatus in a sequence which leads to the loss of a generation of chromatic apparatus.

When meiosis is gametic and fertilisation follows after, the immediate formation of zygote as a cell might contain the maternal as well as paternal centrioles, having endowed as it were, with the extra centriole generation.

According to *Daniel Alazia* the kinetochore is an essential part of the chromosomes so far as the mitosis is concerned. In gametogenesis there is a gain in total chromatic material, since these types of cell behaviour lead to duplication and subsequent reamassing of chromosomes. The extra set of centriole is rejected normally in fertilisation or autogamy. Thus a cell results in a chromatic apparatus. This is accomplished by meiosis : one division meiosis gives the centriole a chance to 'catch up' in a singel duplication.

4

GAMETOGENESIS

The constancy in the number of chromosomes in the body cells of sexually reproducing organisms posed a problem. It should be obvious that in sexually reproducing organisms a reduction in the number would have to occur sometime in life cycle in order to maintain the constant number of chromosomes; if such a reduction did not occur, the number of chromosomes would double with each fertilisation.

The process in which the number of chromosomes reduces is known as *meiosis* and involves two rapid divisions. It occurs in animals in the formation of eggs and sperms and in plants in the formation of spores. The process is essentially the same wherever it occurs in the formation of eggs in animals, in the formation of sperms in animals, and in the formation of spores in plants.

In all of the preceding discussion on the life cycles of both plants and animals, the process of meiosis has been continuously referred to without any discussion of the details of the process. As has been pointed out in above account, the cells in the life cycles of animals are all diploid (2n), with the exception of the eggs and sperms that are haploid (n). The meiotic process in the formation of eggs in animals differs from that in the formation of sperms in one important respect.

In the formation of sperm cells, each cell that undergoes meiosis, forms four functional sperm cells. Each sperm is composed of essentially of a *head* and a tail, which effects e movements of the sperm. In the formation of the eggs, on e other hand, only one functional ovum is formed from each 11 that undergoes meiosis.

GERM CELL FORMATION

The origin of the germ cells during ontogeny, and their differentiation, have long occupied a central position of interest for obvious reasons. In most of higher animals the germs cells seem to arise in the "*germinal epithelium*" which occupies a localised region of the peritoneal or coelomic epithelium. The formation of germ cells can be traced during the early stages of cleavages.

In such cases the primordial germ cells seem to lead an independent life within the soma, almost as if they were parasitic or symbiotic organisms. Zygote is supposed to be the stem cell which gives rise by divisions to both germ cells and somatic cells. From the stem cells arise the primordial germ cells which are then converted into gonia to form gonads. These cells show the same general characters in both sexes, but are sooner or later differentiated as primary *oogonia or spermatogonfa.*

Germ Cell Migration in Amphibians

In anuran amphibians-frogs and toads-the germ plasm is readily marked by granules near the vegetal pole of the fertilised egg. However, this plasm does not remain long in the vegetal region. During cleavage, this material is brought upward through the yolky cytoplasm, and the RNA-rich granules are seen to become associated with the cells lining the floor of the blastocoel.

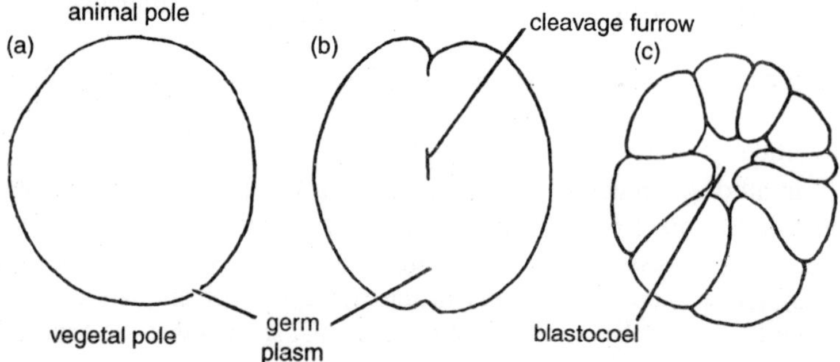

Figure 4.1 : Changes in the position of germ plasm in an early frog embryo. Originally located near the vegetal pole of the uncleaved egg.

Blackler traced the migration of these cells into the gonad by transferring blocks of cells from one strain of *Xenopus laevis* neurulae into another. The primosdial germ cells of this frog are seen to move laterally from the endoderm of the gut to the dorsal mesentery, which connects the gut to the region where the mesodermal organs are forming.

They migrate up this tissue until they reach the developing gonads. *Xenopus* PGCs move by extruding a single filopodium and then streaming their yolky cytoplasm into the filopodium while retracting their tail.

Contact guidance in this migration seems very likely as the cells over which they migrate are oriented in the direction of that migration. Furthermore, PGC adhesion and migration can be inhibited if the mesentary is treated with antibodies against Xenopus fibronectin.

Thus, the pathway for germ cell migration in frogs appears to be composed of fibronectin-coated cells that are oriented in the direction of cell migration. The cells over which the PGCs travel lose this polarity soon after migration has ended.

Intrinsic factors are also important to the specific migration of the primordial germ cells. When the vegetal germ plasm of frog eggs is irradiated with ultraviolet light (which destroys RNA and the disulfide bonds of proteins), the migration of PGCs to the genital ridge is inhibited or delayed significantly. The primordial germ cells of urodele amphibians (salamanders) have a different origin, which has been traced (by reciprocal transplantation experiments) to the regions of the mesoderm that involute through the ventrolateral lips of the blastopore.

Moreover, there does not seem to be any particular localised "germ plasm" in salamander eggs. Rather, the interaction of the dorsal endoderm cells and animal hemisphere cells creates the conditions needed to form germ cells in the particular areas that involute through the ventrolateral s. So in salamanders, the primordial germ cells are already thin the mesodermal region and presumbly do not follow e same path to the gonad.

Germ Cell Migration in Mammals

In mammals, germ cell migration takes a route very similar to that seen in anurans. The major break through in mapping mammalian germ cell migration came when Baxter (1952) demonstrated that mammalian primordial germ cells contained enormously high concentrations of the enzyme alkaline phosphatase. The primordial germ cells could thus be distinguished from other cells by staining for the presence of that enzyme.

The route of the mammalian PGC migration is shown in Figure. The PGCs of mice can first be seen in the 7.5-day-old embryonic allantois. Shortly thereafter, they are seen at the base of the allantois and in the adjacent yolk sac. By this time, they have already split into two populations, which will migrate to either the right or the left genital ridge.

The PGCs then move caudally from the yolk sac through the newly formed hindgut and up the dorsal mesentary into the genital ridge. Most of the PGCs have reached the developing gonad by day 11 postfertilisation.

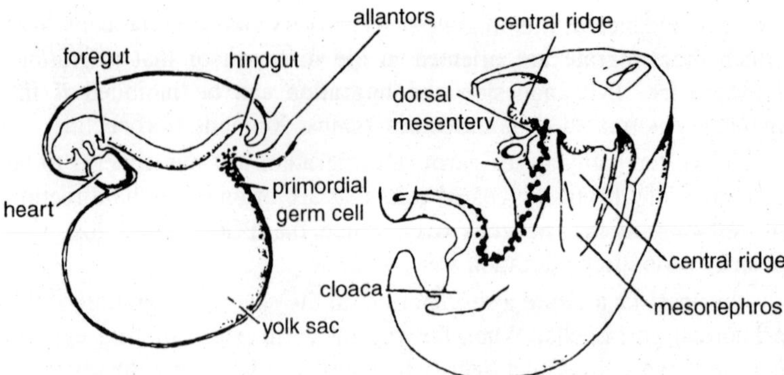

Figure 4.2: Pathway for teh migration of mammalian primordial germ cells. (a) Primordial germ cells first recognised in the yolk sac near the junction of hindgut and allantois. (b) Migration through gut and dorsally up the dorsal mesenery into the genital ridge.

During t trek, they have proliferated such that, whereas only 10-1 cells were seen initially, some 2500-1000 PGCs are in the gonads by day 12. As is the case with PGC migration *Xenopus,* mammalian PGCs appear to be closely associat with the cells over which they migrate, and they move extending filopodia over the underlying cell surfaces.

Germ Cell Migration in Birds and Reptiles

In birds and reptiles, PGCs are derived from epiblastic cells that migrate to a crescent-shaped zone in the endodermal layer at the anterior border of the zone pellucida. This region is called the *germinal crescent,* and the primordial germ cells multiply in this region. Unlike the situation in amphibians and mammals, the primary route for germ cell migration in birds and reptiles is through the blood.

The PGCs enter the blood vessels by *diapedesis a* type of movement that is common to lymphocytes and macrophages and that enables cells to squeeze in and out of small blood vessels. The PGCs thus enter the embryo by being transported in the blood. The PGCs must also "know" to get out of the blood when they reach the developing gonad. When the germinal crescent of a chick embryo is removed, and the circulation of that embryo is joined with that of a normal chick embryo, the primordial germ cells from the normal embryo will migrate into both sets of gonads.

It is not known what causes this attraction for the genital ridges. One possibility is that the developing gonad produces a substance that attracts PGCs and retains them in the capillaries bordering the gonad. [Such substances are known to be secreted by lymphocytes at the sites of infection in order to attract macrophages to that area (chemotaxis) to permit them to pass through the capillary wall (diapedesis).] Another possibility is that the endothelial cells of the gonadal capillaries have a cell surface compound that causes the PGCs to adhere there specifically. Auerbach and Joseph (1984) have shown that several capillary networks can be distinguished from each other on the basis of their cell surfaces and that the endothelial cells from ovarian capillaries differ from all others tested.

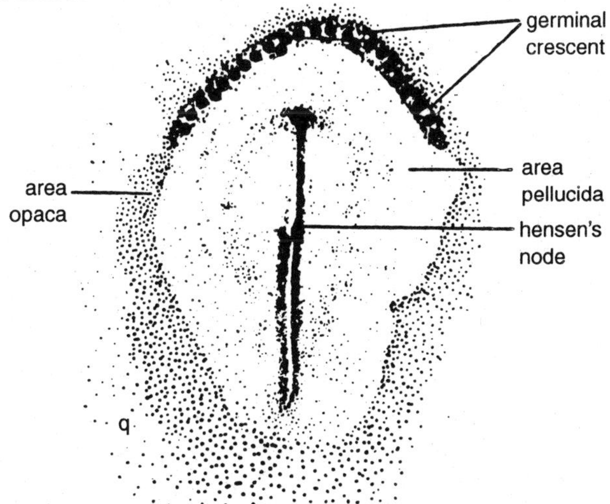

Figure 4.3: Dorsal view of a primitive streak-stage chick embryo showig the region, called the germinal crescent, in which the germ cells arise.

Formation of Sperm

Sperms are formed by the germ cells which are derived 'rom the germinal epithelium of the embryo. The germ cells and somatic cells are always diploid in nature. In higher animals where the definite sex organs are present, the male sex organs are the testes which are composed of a number of germ cells.

The formation of spermamtids from germ cells is called *spermatogenesis* while their metamorphosis into sperm is called *spermiogenesis.* All the germ cells do not take part in spermatogenesis while few of them simply provide nourishment to the cells forming spermatids, these cells are called *nutritive cells* or *Sertolf cells,* while the remaining cells are called the *primary germ cells* or *primordial cells.*

Spermatogenesis

The process of spermatogenesis can be studied under the following

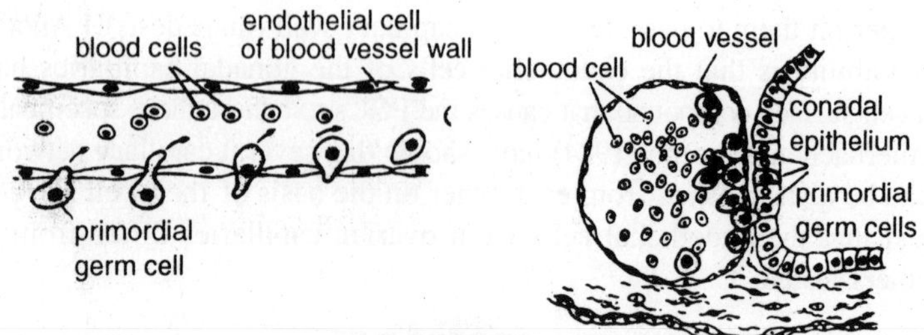

Figure 4.4 : Diapedesis of primordial germ cells in the chick embryo.

heads:

(*a*) Multiplication phase or phase of multiplication.

(*b*) Growth phase or phase of growth.

(*c*) Maturation phase or phase of maturation.

All the germ cells are diploid having a nucleus in the middle of each cell, and in cytoplasm a number of cytoplasmic organ cells are embedded such as Golgi bodies, Mitochondria, Endoplasmic reticulum, Centrosome containing centrioles etc.

(a) Phase of multiplication

As explained above that all the ells of testes do not form spermatids. Those cells which help a nutrition are called the nutritive cells or Sertoli cells and the emaining cells are primordial or primary germ cells. All the primary germ cells divide by repeated mitotic divisions to proluce a number of cells having the same number of chromoomes (diploid). Thus a large number of *spermatogonia* are ormed by the mitotic divisions of primary germ cells.

(b) Phase of growth

After the formation of spermatoal cells, each cell passes in growth phase where the enlarget of cell size as well as nuclear size can be observed. The enlarged cells are called the *primary spermatocytes*. This enlargement in size of cells is due to absorption of nutritive material from the side of cells and also due to the synthetic activities of nucleic acids and proteins. Splitting of chromosomes, linkage and crossing over also take place in this phase for the further divisions of maturation phase.

(c) Maturation phase

After the enlargement of the cell' during phase of growth, the primary spermatocytes jump to the phase of maturation. Here the maturation or reduction division occurs. Reduction division is double as it is composed of two divisions:

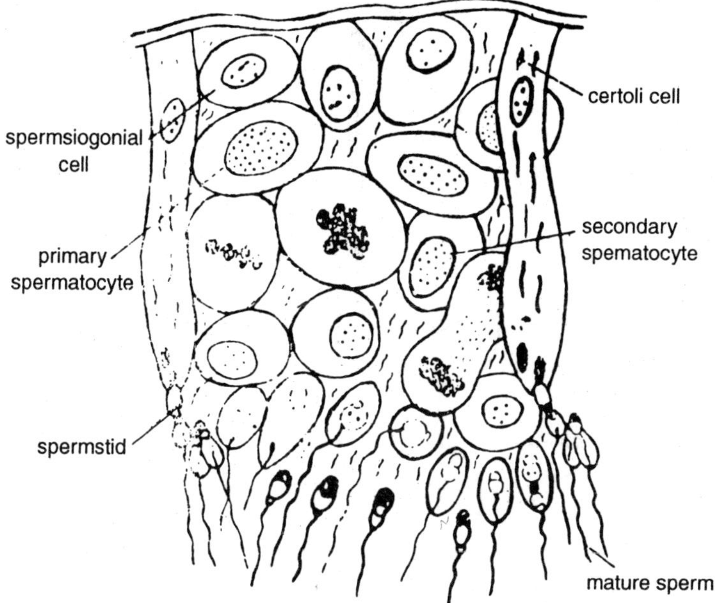

Figure 4.5 : Showing the formation of sperms.

(*i*) Ist meiotic or Ist maturation division which is true reductional in nature.

(*ii*) Second meiotic or second maturation division which is true mitotic in nature.

Two haploid *secondary spermatocytes* are formed from one diploid primary spermatocyte by first meiotic division, and from each haploid secondary spermatocytes, two haploid *spermatids* are formed by the second meiotic division.

Thus from one primary diploid spermatocyte, four haploid spermatids are formed during the phase of maturation. The spermatids do not divide further, so after their formation from germ cells, they undergo metamorphosis to form sperms.

Metamorphosis or Spermiogenesis

During spermiogenesis a number of haploid spherical, spermatids are formed. Each spermatid contains a haploid nucleus, some what in the middle and is surrounded by the cytoplasm and plasma membrane. In the cytoplasm are embedded a number of cytoplasmic organelles such as Golgi bodies, Mitochondria, Centrosome etc.

These various organelles undergo certain modification during the

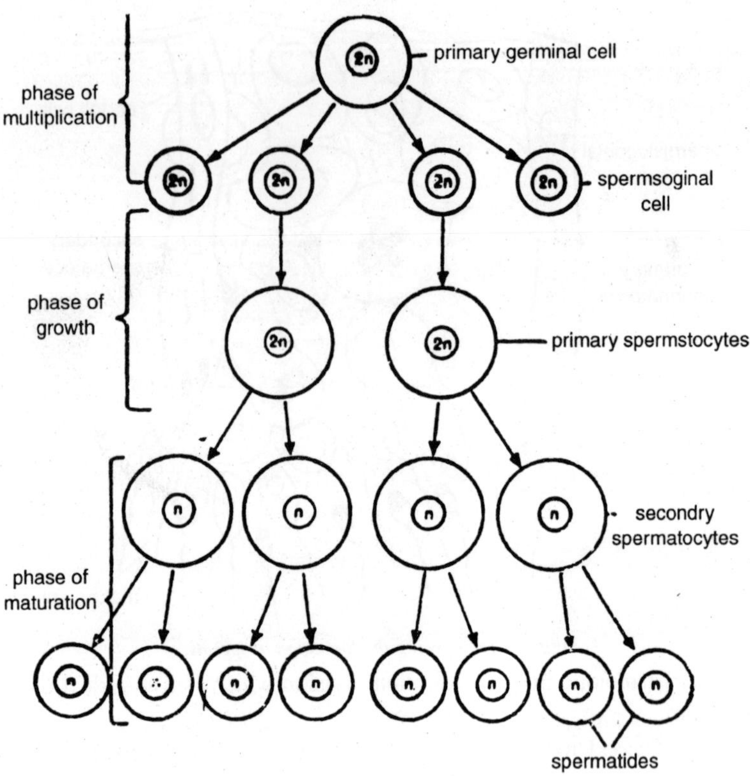

Figure 4.6 : Diagramatic representation of spermatogenesis.

process when spermatids form the spermatozoa. These changes are well studied in Hemipterans.

Role of Centriole during Spermiogenesis

Centrosome is generally present in the vicinity of nucleus towards the anterior side. Centrosome may have one or two centrioles. If there is one centriole, it divides during the process to form two centrioles. Then the membreane encircling treh centrioles, breaks down. Thus bothe the centrioles beocme free lying in the cytoplasm.

One of these centrioles remains against the nuclear membrane and is called the *proximal centriole* while the other one which is away from nuclear embrane is called the distal centriole. From the *distal centriole* a delicate filament which is *axial filament* emerges out piercing the plasma membrane. After the formation of *axial filament*, both the centrioles along with axial filament move towards the posterior side of the nucleus, where they take their final position.

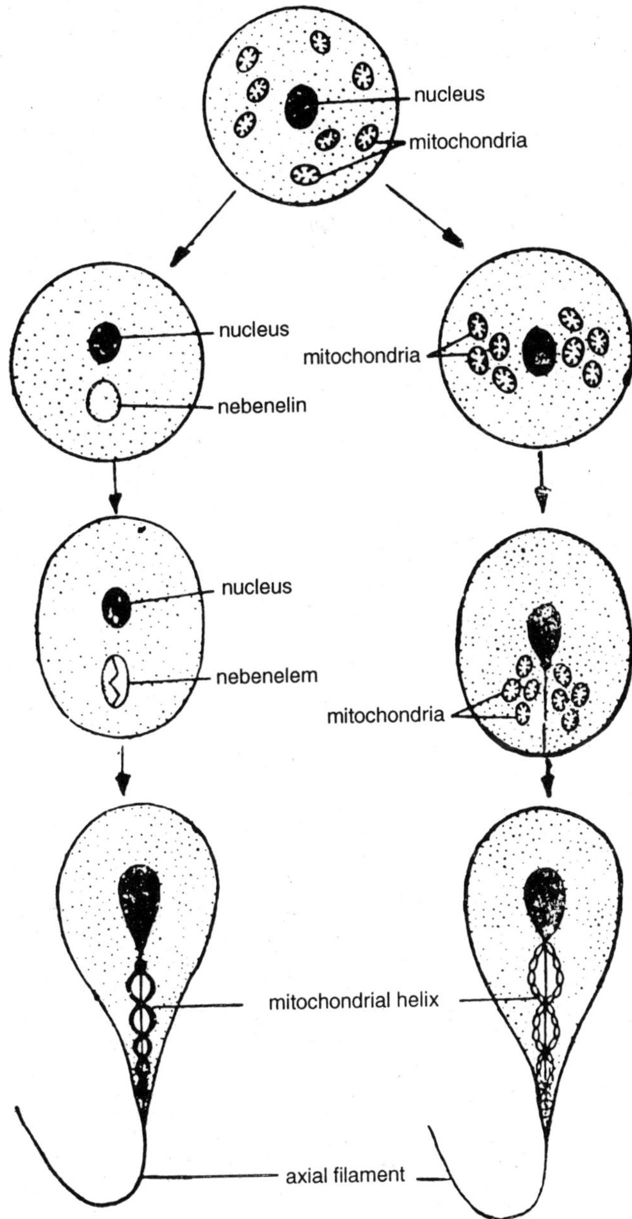

Figure 4.7 : Mitochondria in spermiogeneals.

Here the *distal centrole* which may form a ring divides into two, one of which forming the *end knob* and the other passing back as a ring to the end of the connecting piece. The axial filament, when viewed under the ordinary microscope, is seem to be composed of single fibre but when seen by electron microscope eleven fibres can be observed easily, arranged in two rings, out of which two fibrils are single and are present in themiddle while remaining nine double fibrils are arranged against the periphery. But in mammalian sperm one additional ring of nine single fibres are present just out side the ring of nine tubular doublets.

Role of Mitochondria in Spermiogenesis

The helicular portion of the collar region of the sperm is formed by the mitochondria of spermatid. This helicular portion is called the *mitochondrial helix* or *spiral*. The majority of workers believe that the mitochondrial spiral is a continuous structure formed by the fusion of mitochondria end to end. But according to *Nath* (1965) all the mitochondria of spermatid do not fuse but retain their individuality. The formation of mitochondrial helix is well studied in insects where all the mitochondria form a more or less compact body known as *nebenkern*.

Soon after its formation, the inner contents of the nehenkern is differentiated into dark stained portion or areas forming a series of concentric plates alternating with layers of the lighter substance. Then the nebenkern elongates in length and at the same time a furrow is developed which moves inwards and eventually cuts the nebenkern into two portions.

Both the portions constitute two elongated ribbon like structures having lighter and denser portions respectively. Then both these filaments coil around each other but in such a precise manner that the axi 1 filament is covered on either side by the coiling of filaments which now at this stage constitutes the mitochondrial helix or spiral.

According to Burgos and Fawcett (1955)

During the spermiogenesis the internal structure of mitochondria undergoes a marked change, the cristae which are perpendicular to the outer membrane in normal mitochondria become irregular in outline and may fill the internal canal of spiral as in opossum while *Yasnzumi* (1956) has reported that in mitochondrial spiral of albino rat, the cristae are entirely absent.

Role of Golgi bodies in Spermiogenesis

The acrosome of the sperm is developed from the material of

Golgi complex of the spermatid. During spermiogenesis the entire material of Golgi complex, which is now called dictyosomes accumulate at a place somewhat near the posterior region of the nucleus. Then the dictyosomes fuse to form a single large acroblast.

The acroblast soon begins to migrate towards the other or anterior side of the nucleus. During the migration of acroblast, a granule is secreted by the wall of acroblast which is acrosomal granule. The size of this granule gradually increases, and when the acroblast reaches the anterior side of the nucleus to take its final position, the size of acrosomal granule reaches its peak and fills the entire space of acroblast.

At this stage the term acrosome can be applied easily for the acrosomal granule. Besides the acrosome, the remaining portion of Golgi complex is called the Golgi remnant, passes backward along the tail and ultimately cast out of the cell at the close of spermiogenesis. The acrosome after reaching its definite position near the anterior end undergoes a further differentiation, within it, appears a deeply staining mass, which rapidly develops into the pointed perferotorium at the tip of the spermatozoon.

Role of Nucleus in Spermiogenesis

No conspicuous changes occur within the nucleus until the acroblast and its differentiated acrosome complete their migration around it. The chromatin substance then arrange itself in a layer against the inside of nuclear membrane. Then the chromatic portion of the nucleus elongates and begins its transformation into the slender head of the sperm.

The bead continues to elongate in the form of a narrow spoon with the acrosomic band lying in the hollow; eventually it becomes a slender pointed body in which nucleus and acrosome can no longer be distinguished. During these changes non-chromatic constituents of nucleus gradually diminishes and disappears so that the nucleus of spermatozoon is an extreme concentrated chromatic mass in which structures can not be demonstrated.

As the other components of the spermatid undergo the transformation, the cytoplasmic cell body elongates and remains visible around them until a comparatively late stage. Towards the close of spermiogenesis, in insects, it has been ascertained that a cytoplasmic ball containing Golgi remnant and other granules of unknown nature is sloughed off and ingested by the nutritive epithelial cells lining the spermatocyte. In other animals a cptoplasmic mass is like wise eleminated but the amount of active protoplasmic contained in it, is uncertain.

STRUCTURE OF A MATURE SPERM

The body of a mature sperm can be dividhd into following parts:

Head

The head is the uppermost portion of the sperm bearing acrosome developed by Golgi bodies. It may be spheroidal as in teleosts and certain invertebrates, rod shaped or filliform as in insects; flattened or spoon shaped as in Guinea pig, man and many other mammals; book shaped as in rat etc. Head f the sperm is further differentiated into two parts:

(i) Acrosomal part

The anterior most region of head which is smaller than the nuclear part is called the acrosomal part. It exhibits a great diversity of form; varying from a small granule or knob to a large conical or spine like process.

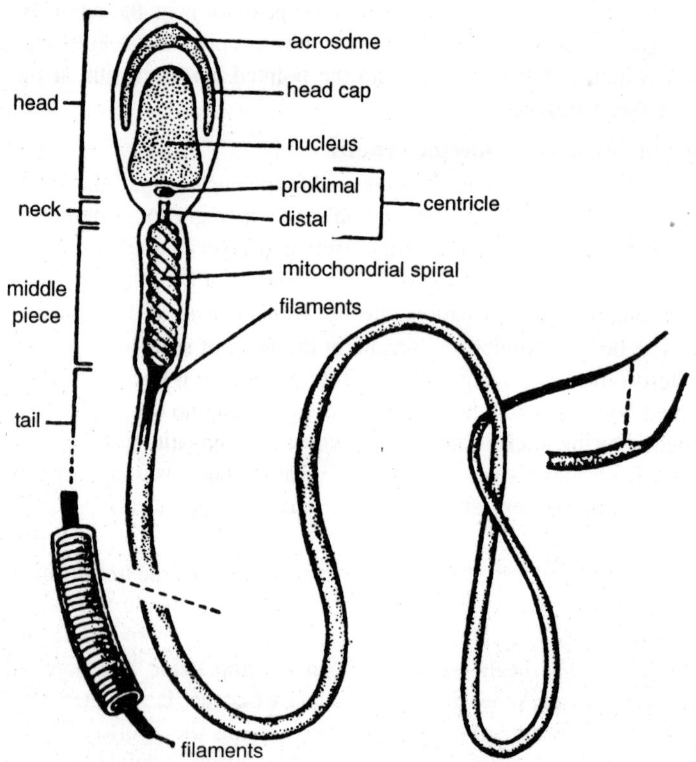

Figure 4.8 : Structure of a mature sperm.

The acrosome of acrosomal part is developed from the Golgis bodies of the spermatid.

(ii) Nuclear part

Constitutes of nucleus and occupies thenajor portion of head. Nucleus of sperm is composed of nainly a solid and homogenous mass of basichromatin. The development of the sperm shows that both nucleus and acrosome are covered by a very thin cytoplasmic covering, perhaps a cell membrane.

Neck

It is the portion of sperm which is present between head and collar region. It forms a stalk by which the flagellum is ittacbed to the head. This region often appears not to be surrounded by a mitochondrial sheath. Products of proximal centriole are lying in this region. The relations of the centriolar apparatus to the neck region are of great interest because of the evidence that from this region of the sperm, or in its neighborhood arises the sperm center during the fertilisation of the egg.

Neck region is not distinct in mammalian sperm, where it is distinct, it marks the beginning of the axial filament.

The Middle piece

It is the portion of sperm where the axial filament is covered by the helicular structure of mitochondria. Mitochondria) helix is connected by the products of distal centrioles on either ends. This mitochondria) helix provides energy for the movement of sperm during fertilisation. The energy resides in mitochondrial helix in the form of ATP.

Tail

Generally tail is present in all the sperms which is in the form of an elongated filament (axial filament) and is extruded out from the distal centriole during spermiogenesis. But in certain cases as in nematodes and arthropods (myripods, arachnids, crustaceans etc.) the sperms are not provided with tail.

Tail is further differentiated into two parts or pieces:

(i) Main piece

It is that portion of axial filament or tail which is enveloped by cytoplasmic sheath. This sheath may either cover the major portion of the filament or the entire tail may be sheathed as in urodela.

(ii) End piece

The portion of tail or axial filament which is not envelope by

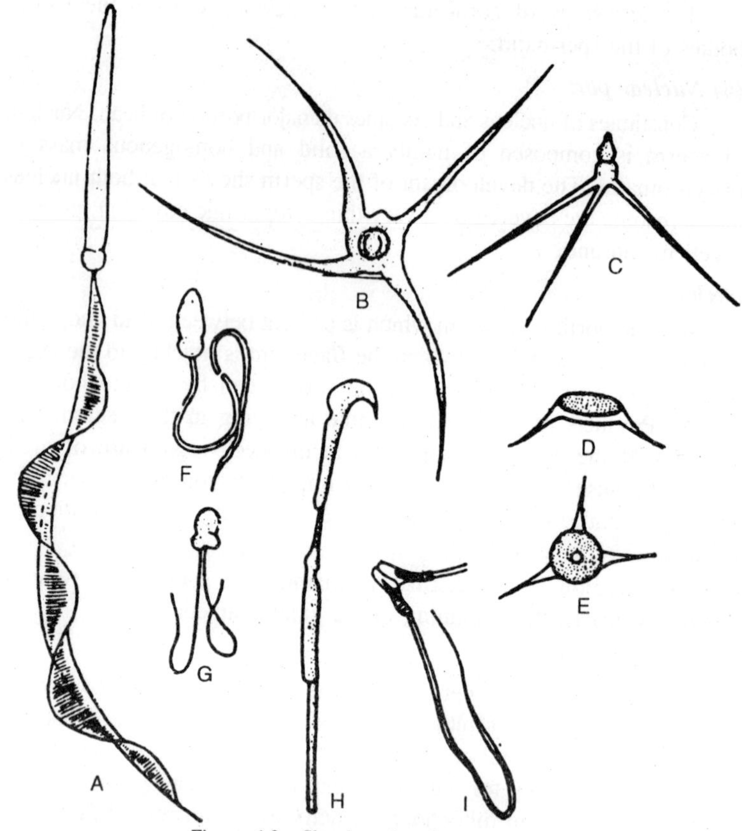

Figure 4.9 : Sketches of various sperms.

cytoplasmic sheath, constitutes the end piece, In Urodela, as the entire tail is sheathed, end piece is lacking.

(*a*) Toad ; (*b*) Cray fish ; (*c*) Ethusa ; (*d*) (*e*) Crab ; (*f*) Sea urchin ; (*g*) Toad fish ; (*h*) Guinea pig ; (*i*) Opossum.

Gene expression during sperm development

Gene transcription during spermatogenesis takes place predominantly during the diplotene stage of meiotic prophase. This has been observed in many organisms, but the best-documented case is probably that of Y chromosome transcription in *Drosophila hydel.* Here, RNA transcripts originating from the Y chromosome are seen to be essential for controlling spermiogenesis.

When we recall the function of the Y chromosome in *Drosophila,* this is not surprising, for the Y chromosome is not involved in sex determination here. Rather, it is needed for the formation of viable sperm. The difference between XY *Drosophila* and XO *Drosophila* is that the latter are sterile. Both are male. In *Drosophila hydei,* the Y chromosome extends five prominent loops of DNA. If any of these loops is deleted, the organisation of the sperm tail will be abnormal.

All the component parts are present, but they are not organised properly. It appears, then, that Y-specific RNA made during meiotic prophase is utilised later during spermiogenesis. This storage of RNA made during the diplotene stage has been seen elsewhere. In mammals and birds, a specific form of lactate dehydrogenase, LDH-X, is made during spermatogenesis.

This protein enables the developing sperm to utilize pyruvic acid as an alternate energy source. The message for this protein can be identified in spermatocyte cytoplasm during meiotic prophase. Similarly, in many species, a small protein called Protamine appears in the nucleus during the final stages of spermiogenesis. This protein contains about 32 amino acids of which all but four or five are arginine residues. This protein replaces the nuclear histones and causes the DNA to form a compact, almost crystalline, array.

DNA complementary to trout protamine mRNA can detect protamine message sequences in the primary sp:rmatocyte. However, these messages are stored in ribonucleoprotein particles. Only during the spermatid stage, approximately a month after its synthesis, is the protamine message transcribed into protein.

We find, then, that just like the oocyte, which can synthesize RNA in the diplotene stage and store it for later use, the sperm can also package messages for later translation. In addition to gene transcription during meiotic prophase, there is also evidence that certain genes are transcribed in thespermatids. This evidence for Haploid Gene Expression comes from studies involving heterozygous mice in which two different populations of sperm are seen ro exist—one population expressing the mutant phenotype and one population expn sing the wild-type trait.

If the synthesis of the RNA or prote were to occur while the cells were still diploid, all the sper would show the same phenotype. One such mutation for which two sperm populations can be separated from heterozygotes is the t^{12} mutation in mice. This mutation, as we have earlier, is lethal when homozygous The morulae do not undergo compaction.

When heterozygous, however, this mutation leads to a phenomenon called Segregation Distortion. Heterozygous ($t^{12}/+$) males mated to

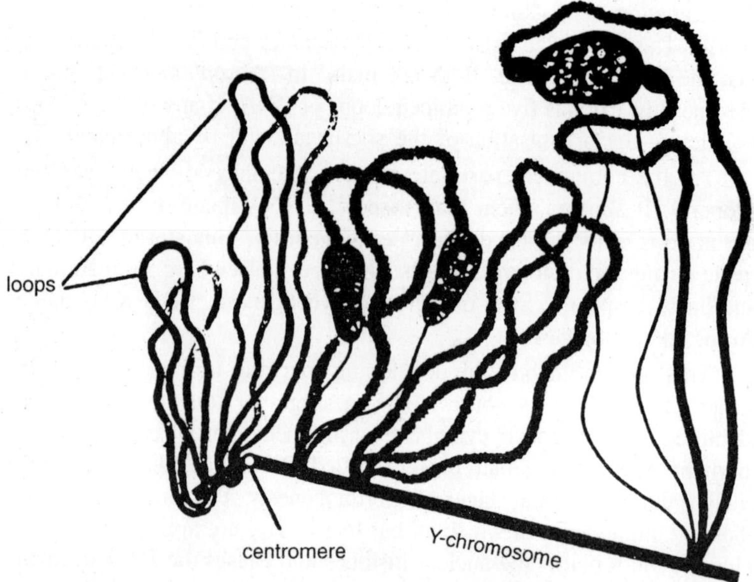

Figure 4.10 : Diagram of the Y chromosome of Drosophila hydel in the lambrush chromosome state. Five loops are readily observed.

wild-type (+/+) females will produce heterozygous offspring over 95 per cent of the time, instead of the expected 50 per cent. The reciprocal crossing of wild-type males and heterozygous females does give the expected Mendelian 1 : 1 ratio. It appears, then, that the t^{12}-bearing sperm differed from the sperm carrying the wild-type allele.

This was confirmed by immunological analysis. It was known that the product of the wild-type allele of t^{12} is present on the cell surface of the sperm. When antibody against this wild-type product was added to sperm from a heterozygous mouse, it bound to roughly 50 per cent of them. This indicates that half the sperm expressed the wild-type allele and half did not. Therefore, the transcription of the t^{12} gene occurs after meiosis, rather than before it. Transcription is occurring from the baploid genome.

OOGENEIS

Oogenesis is the process in which eggs or ova are produced from the germ cells which are derived from the germinal epithelium of embryo. The entire process of oogenesis can be divided into three phases or stages

(*i*) Phase of multiplication.

(*ii*) Phase of growth.

(*iii*) Phase of maturation.

(i) Phase of multiplication

During this phase primary germs cells divide in such a way that chromosome number in daughter cells remains as such as they (germs cells) divide mitotically to produce a large number of daughter cells which are the *oogonial cells*. Oogonial cells contain the diploid number of chromosomes as they are produced from the diploid primary germ cells by mitotic divisions.

(ii) Phase of growth

Phase of multiplication is replaced by growth phase, where the oogonial cells simply enlarge in size due to the absorption of nutritive material from the surrounding medium and also due to the synthesis of nucleic acids and proteins. The phase of growth can also be labelled as Ist prophase of reduction division as the replication of chromosomes, linkage and crossing over occur during this phase.

After these Changes each oogonial cell is then called the *primary oocyte*. *All* the primary oocytes are once again diploid as none of the Jivision has been traced during this phase, so there is no dis:urbance in the number of chromosomes.

(iii) Phase of Maturation

The phase of growth is replaced)y phase of maturation. During this phase the reduction livision occurs which is double in nature, at it is composed)f first meiotic division which is true reduction division and econd meiotic division which is true mitotic in nature.

But in oogenesis both the divisions are incomplete as in both the cases nucleus divides properly but the division of cytoplasm is not regular. In first meiotic division the diploid nucleus of primary oocyte divides to form two haploid nuclei, but the cytoplasm does not divide. Then one of the daughter nucleus moves towards the periphery and penetrates out of the vitelline membrane to lie beneath the second egg membrane, the chorion.

This migrated body constitutes the *first polar body* while the remaining structure with haploid nucleus is called *secondary oocyte*. Then the haploid nucleus of secondary oocyte divides by second meiotic division to form two daughter haploid nuclei. Again one of these nuclei migrates towards the periphery to constitute the *second polar body*. After the extrusion of second polar body the developing structure is called the *ootid*. The first polar body may or may not divide to form two polar bodies. The polar bodies disintegrate later on. Here in oogenesis only one haploid ootid is formed from one diploid oogonial cell.

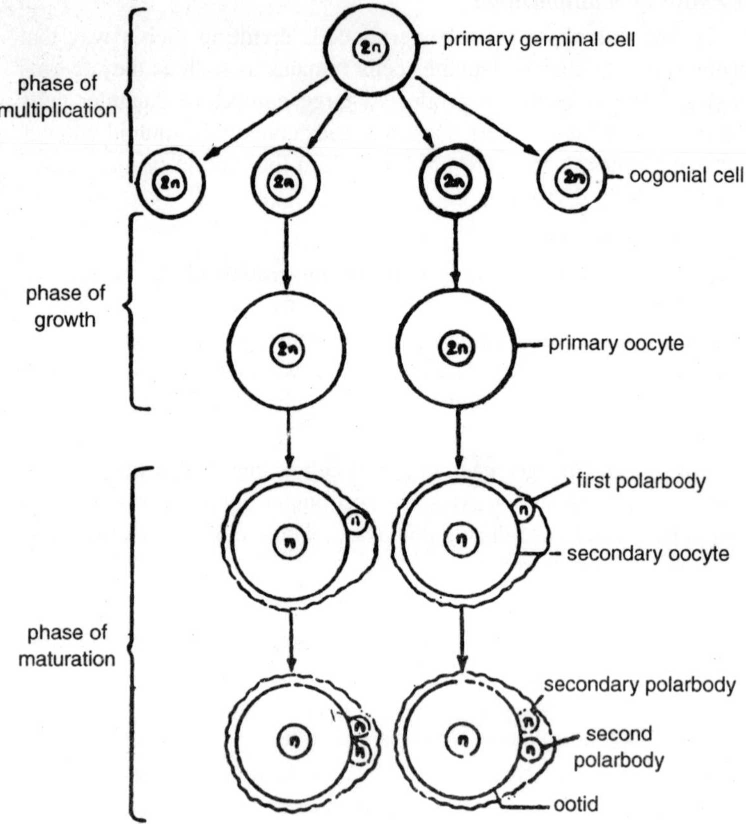

Figure 4.11: Diagramatic representation of oogenesis.

Changes in the Cytoplasm

An early oocyte is spherical in outline having a nucleus in the middle, while the remaining portion of the oocyte is filled with cytoplasm alongwith a number of cytoplasmic organelles such as Mitochondria, Golgi bodies, cell center etc. All these organelles take an important role in *vitellogenesis.*

Cytocenrum or cell center or centrosome is situated in the vicinity of nucleus enclosing one or two centrioles. The word *archoplasm* or *idiosome* can be applied for the denser cytoplasm in which centrioles are embedded in centrosome. Then soon the nucleus which is haploid in nature is surrounded by rings or circles developed by the arrangement of Golgi bodies, mitochondria and centrosome in circles. The entire circular rings thus formed can be labelled as *Balbianic Vitelline rings* or *Yolk nucleus. Yolk* nucleus may either surround the original nucleus of the developing ovum containing chromosomes, or may lie adjacent to it.

The presence of yolk nucleus can be observed clearly in young oocyte. Then the entire organelles of yolk nucleus such as Golgi bodies, mitochondria, centrosome etc., disperse through the cytoplasm and disappear leaving a number of granules at their places. These granules or globules are the *yolk granules* or *yolk globules.* (The detailed account of yolk formation has been given under separate heading.) The yolk thus formed may be of any of the following types:

a) Protein yolk.

b) Carbohydrate yolk.

c) Fatty yolk.

On the basis of amount and presence or absence of yolk particles the eggs of animals can be classified as follows:

1. *Alecithal.* The eggs without the yolk are known as alecithal eggs, e.g., Eutherian mammals.

2. *Microlecithal.* The eggs with very little amount of yolk are known as the microlecithal eggs, e.g., *Amphioxus.*

3. *Macrolecithal.* The eggs with large amount of the yolk are known as the macrolecithal eggs. The macrolecithal eggs can be grouped into two types on the basis of the distribution of the yolk in the ooplasm

A. Homolecithal. The eggs which evenly distributed yolk con cents in the ooplasm are know as the homolecithal eggs, e.g , eggs of echinoderms.

B. Heterolecithal. The eggs in which the yolk is not evenly distributed in the ooplasin are known as the heterolecithal eggs. The heterolecithal eggs may be of following types

(i) Telolecithal. When the amount of the yolk is concen trated in the one-half of the egg to form thethe egg, then this condition is known as the telolecithal, e.g., eggs of amphbians, etc.

(ii) Meiolecithal. In the meiolecithal eggs the amount of the yolk is very large and it occupies the largest portion of the egg except a small disc-shaped portion of the cytoplasm. The cytoplasm contains the zygote nucleus and is known as the germinal disc. e.g., eggs of fishes, reptiles and birds.

(*iii*) *Centrolecithal.* In the centrolecithal eggs the yolk accumulates in the centre of the ooplasm, e.g., eggs of insects.

Changes in the Nucleus. No consipicuous changes have been traced upto the formation of primary oocyte. After its formation a sequence of changes have been observed. When the oogonial cells enlarge in size to enter in the *phase of growth* or *vegetative phase* and form the primary oocytes, the chromatin material assumes definite chromosomes. The free ends of all the chromosomes then arrange themselves in such a precise manner that they face towards one pole.

This stage is called the *bouquet stage.* The chromosomal threads which constitute the bouquet arrangement are also known as *leptotene threads.* The leptotene threads now conjugate to give rise haploid number of bivalents. At the same time numerous side branches or side loops are developed by the chromosome. Such chromosomes bearing side loops are called *lamp brush chromosome.*

Nucleolus also plays an important role in the synthetic ~tivity taking place in the nucleus. The small nucleus may ve rise to several nucleoli by fragmentation and budding. All these nucleoli come to lie near the nuclear membrane.

Due to these changes the outline of nucleus becomes ir,gular and shows sac like protrusins through which nucleolus cortercts pass out.

The maturation of the oocyte in amphibians

The egg is responsible for initiating and directing development, and in some species (as seen earlier) fertilisation is not even necessary. The accumulated material in the oocyte cytoplasm includes energy sources and organelles (the yolk and mitochondria), the enzymes and precursors for DNA, RNA, and protein syntheses, stored messenger RNAs, structural proteins, and morphogenic determinants.

A partial catalog of the materials stored in the oocyte cytoplasm is shown in Table elsewhere in this chpater. Most of this accumulation takes place during meiotic prophase I, and this stage is often subdivided into Previtellogenic (Greek, meaning "before yolk information") and Vitellogenic (yolk-forming) phases. Eggs of fishes and amphibians are derived from a stem-cell oogonial population, which can generate a new cohort of oocytes each year. In the frog *Rana pipiens,* oogenesis lasts 3 years. During the first 2 years, the oocyte increases its size very gradually. During the third year, however, the rapid accumulation of yolk material in the oocytes causes them to swell to their characteristically large size. Eggs mature in yearly batches, the first cohort maturing shortly after metamorphosis; the next group matures a year later.

Table 4.1 : Cellular components stored in the mature oocyte of *Xenopus laevls.*

Component	*Approximate excess overamount in larval cells*
Mitochondria	100,000
RNA polymerases	60,000-100,000
DNA polymerases	100,000
Ribosomes	20,000
tRNA	10,000
Histones	15,000
Deoxyribonucleoside triphosphates	2,500

Vitellogenesis occurs when the oocyte reaches the diplotene stage of meiotic prophase. This is also the stage when the lampbrush chromosomes of the nucleus are seen to be actively synthesizing RNA. Yolk is not a single substance, but a mixture of materials used for embryonic nutrition. The major yolk component is a 470.000-*d* protein called Vitellogenin.

It is not made in the oocyte (as are the major yolk proteins of such organisms as annelids and crayfish), but it is synthesized in the liver and carried through the blood to the ovary. This enormous protein passes between the follicle cells of the ovary and is incorporated into the oocyte by Micropinocytosis, the pinching off of membrane-bounded vesicles at the base of microvilli Dumont (1978).

In the mature oocyte, vitellogenin is split into two smaller proteins, the heavily phosphorylated Phosvitin and the lipoprotein Lipovitellin. These two proteins are packaged together into membrane-bounded Yolk Platelets. Glycogen granules and lipochondrial inclusions store the carbohydrate and lipid components of the yolk, respectively.

As the yolk is being deposited, the cortical granules begin to form from the Golgi apparatus. They are originally scattered randomly through the oocyte cytoplasm but then migrate to the periphery of the cell. The mitochondria are capable of self-replication, and they divide to form millions of organelles.

These mitochondria will be apportioned to the different cells during cleavage, and new mitochondria are not formed until after gastrulation is initiated. As vitellogenesis nears an ends, the oocyte cytoplasm becomes stratified. The cortical granules, mitochondria, and pigment granules are found at the periphery of the cell, creating the oocytes cortex. Within the inner cytoplasm, distinct gradients emerge. The yolk platelets are found more heavily concentrated at the vegetal pole of the oocyte, whereas the glycogen granules, ribosomes, lipochondria, and endoplasmic reticulum are found more towards the animal pole. The mechanisms for establishing these gradients are not known.

In *Xenopus,* the leptotene stage of meiosis lasts only 3-7 sygotene takes from 5-9 days, and pachytene persists roughly 3 weeks. The diplotene stage, however, can last years. Even so, vitellogenesis occurs only in part of the diplotene, and the signal for the breakdown of the germinal vesicle occurs after vitellogenesis is completed.

The regulation of these events is controlled by the hormonal interactions between the hypothalamus, pituitary gland, and follicle cells of the ovary. When the hypothalamus receives the cues that the mating season has arrived, it secretes gonadotropin releasing hormone, which is received by the pituitary.

The pituitary responds by secreting the gonadotropic hormones into the blood. These hormones stimulate the follicle cells to secrete estrogen, and the estrogen instructs the liver to syntnesize and secrete vitellogenin. Upon estrogen stimulation, the liver cells change drastically. These changes are usually brought about during the annual mating season, but when estrogen is injected into adult frogs (either male or female) at any time, these changes can be induced and vitellogenin secreted. Estrogen induces vitellogenin at both the transcriptional and translational levels. Before estrogen, there are no detectable vitellogenin messenger RNAs in the liver.

Following hormone administration, each cell contains some 50,000 vitellogenin mRNA molecules, constituting roughly half the total cellular mRNA. The transcriptional activation of the vitello genin genes would normally generate only about 1500 of these mRNA molecules per cell. However, estrogen also specifically stabilizes that message such that its half-life increases from 16 hours to 3 weeks. Because these effects occur in the absence of new protein synthesis, they appear to be direct consequences of estrogen.

Thus, estrogen is seen to control the accumulation of vitellogenin by transcriptional and translational gene regulation. Release of the primary oocyte from meiotic arrest requires progesterone. This hormone is secreted by follicle cells in response to gonadotropic hormones secreted by the pituitary gland. Within 6 hours of progesterone stimulation Germinal Vesicle Breakdown (GVBD) occurs, the microvilli retract, th. nucleoli disintegrate. and the lampbrush chromosomes contract and migrate to the animal pole to begin division. Soon afterward, the first meiotic division occurs, and the mature ovum is released from the ovary by a process called Ovulation.

How does progesterone accomplish this result ? First, it is thought that progesterone works at the surface of the oocyte because microinjection of progesterone directly into the oocyte cytoplasm does not cause GVBD. Second, progesterone is thought to cause the release of free calcium ions, which had formerly been bound within the cortext. This increase is observed within a minute of progesterone administration, and GVBD can be induced by the injection of calcium ions directly into the oocyte or by incubating the oocytes in a solution containing the calcium ionophore A23187.

The release of free calcium ions has been correlated with a decline in cyclic AMP (cAMP) levels within the oocyte. Because the addition of cAMP to oocytes inhibits GVBD, it seems likely that the increase in calcium causes GVBD by causing a decline in cAMP levels. Cyclic AMP usually acts by activating certain enzymes, which phosphorylate other proteins. Without cAMP, then, certain proteins would not be phosphorylated. It is hypothesized that one of these proteins is an active "initiator protein."

This protein would be inactivated by phosphorylation. When the cyclic AMP levels decline, this protein is not phosphorylated, and it is therefore active. It can then catalyze a reaction (probably a cAMP-independent phosphorylation) that activates a formerly inactive protein called Maturation Promoting Factor (MPF).

This MPF is known to exist because cytoplasm from mature oocytes will cause GVBD when injected into immature oocytes. Thus, progesterone appears to activate a cascade of post-translational modification steps that result in the activation of MPF. The mechanism by which MPF initiates germinal vesicle breakdown and the other steps of maturation is still unknown. Another effect of progesterone is to shut down RNA synthesis.

Unlike the situation in sea urchin oocytes, in which the amount and the complexity of the mRNA population are both seen to increase with oocyte maturation, the amount and the complexity of frog oocyte mRNA appears to say relatively constant. The amount of increase seems to be balanced by that of degradation, even during the lampbrush stage. Progesterone also initiates the translation of stored messenger RNAs in the oocyte cytoplasm. Thus, whereas in sea urchins the translation of maternal messages is initiated by fertilisation, here the signal for such translation is initiated by progesterone as the egg is about to be ovulated.

Oogenesis in meroistic insects

Certain insects undergo Meroistic oogenesis wherein cytoplasmic connections remain between the cells produced by the Dogonium. In *Drosophila,* each oogonium divides four times to produce a clone of 16 cells connected to each other through Ring Canals. The production of these interconnected cells (called Cystoc)tes) involves a highly ordered array of cell division. Only those two cells having four interconnection are capable of developing into oocytes, and of these two, only one prevails.

The other begins meiosis but does not complete it. Thus, only one of the 16 cystocytes can become an ovum. All the other cells become Nurse Cells. As it turns out, the cell destined to become the oocyte is that cell residing at the most posterior tip of the egg chamber that encloses the 16-cell clone. The oocytes of meroistic insects do not pass through a transcriptionally active stage, nor do they have lampbrush chromosomes.

Rather, autoradiographic evidence shows that RNA synthesis is largely confined to the nurse cells and that the RNA made by these cells is actively transported into the oocyte cytoplasm. When the egg chambers of a housefly are incubated in radioactive cytidine, the nuclei of the nurse cells show intense labeling. When the labeling is stopped and the cells are incubated for 5 more hours in non-radioactive media, the labeled RNA is seen to enter the oocyte from the nurse cells.

Oogenesis takes place in only 12 hours, so the nurse cells are extremely busy during this time. They are aided in their transcriptional efficiency by becoming polytene. Instead of having two copies of each chromosome, they replicate their chromosomes so as to produce 512 copies. The 15 nurse cells have been seen to transport both ribosomal and messenger RNAs into the oocyte cytoplasm, and entire ribosomes may be transported as well.

The mRNAs do not associate with poly;omes, a finding that suggests that they are not immediately fictive in protein synthesis. The meroistic ovary confronts us with some interesting problems. If all the cells are connected such that proteins and RNAs shuttle freely between them, why should they have different developmental fates ? Why should one cell become the oocyte while the others become "RNA synthesizing factories"? Why is the flow of protein and RNA in one direction only?

One answer to these problems is that within the egg chamber an electrical gradient exists and prevents the diffusion of certain molecules while facilitating the diffusion of others. Woodruff and Telfer (1980) have shown that such a gradient exists and that it inhibits the diffusion of most proteins while enhancing the diffusion of acidic macromolecules. When an acidic enzyme is labeled with a fluorescent marker and injected into nurse cells, the protein is seen to cross from the nurse cell into the oocyte cytoplasm.

However, when that same labeled protein is made basic by small chemical modifications, it will remain in the nurse cell into which it was injected. In this way, levelopmental cues are not exchanged between the connected -ells, but certain proteins and nucleic acids can be transported in a unidirectional fashion. The three major yolk proteins in *Drosophila* are made in the at body and ovary, but not in the oocyte itself.

The hormotal control of yolk synthesis is controlled by juvenile hormone, cdysone, and a neurosecretory hormone from the brain. It is thought that the brain hormone, responding to environmental cues, stimulates the corpora allata to secret juvenile hormone. Juvenile hormone (1) regulates the nptake of the yolk peptides at the surface of the oocyte, (2) stimulates the synthesis of ovarian yolk proteins (which are each identical to those made by the fat body, and (3) causes the ovarian follicles and other abdominal cells to secrete ecdysone. The ecdysone stimulates the fat body to produce yolk proteins just as estradiol stimulated the amphibian liver to do so.

Similarly, the administration of ecdysone to adult males will cause their fat bodies to secrete yolk proteins as well. However, whereas the amphibian oocyte was seen to be an extraordinary active cell, we see here an oocyte that is essentially passive and that is supplied with the necessary cytoplasmic components by its harder working neighbours.

Oogenesis in humans

The maturation and ovulation of the mammalian egg follows one of two basic patterns, depending upon the species. One type of ovulation is stimulated by the physical act of intercourse itself. Physical stimulation of the cervix triggers the release of gonadotropins from the pituitary. These gonadotropins signal the egg to resume meiosis and initiate the events expelling the ovum from the ovary.

This method ensures that most copulations lead to fertilised ova, and animals that utilise this method of ovulation—rabbits and minks-have a reputation of procreative success. Most mammals, however, have a periodic type of ovulation. The female ovulates only at specific times of the year, called Estrus (or its English equivalent, "beat"). In these cases, environmental cues, most notably the amount and type of light during the day, stimulate the hypothalamus to release gonadotropin releasing factor. This factor stimulates the pituitary to release its gonadotropins-follicle stimulating hormone (FSH) and luteinzing hormone (LH)-which cause the follicle cells to grow and to secrete estrogen. The estrogen subsequently binds to certain neurons and evokes the pattern of mating behaviour characteristic of the species. Gonadotropins also stimulate follicular growth and the initiation of ovulation. Thus, estrus and ovulation occur closely together.

Humans have a variation of the theme of periodic ovulation. Although human females have a cyclic ovulation and no definitive yearly estrus, most of human reproductive physiology is shared with other primates. The characteristics primate periodicity in maturing and releasing ova is called the Menstrual.

Cycle as it entails the periodic shedding of blood and cellular debris from the uterus. The menstrual cycle represents the integration of three very different activities : (1) the ovarism cycle, the function of which is to mature and release an oocyte ; (2) the uterine cycle, the function of which is to provide the appropriate environment for the developing blastocyst to implant ; and (3) the cervical cycle, the function of which is to allow sperm to enter the female reproductive tract only at the appropriate time. These three functions are integrated through the hormones of the pituitary, hypothalamus, and ovary.

The majority of the oocytes within the adult human ovary are arrested in the prolonged diplotene stage of the first meiotic prophase (often referred to as the dictyate state). Each oocyte is enveloped by a Primordial Follicle consisting of a single layer of epithelial Granulosa cells and a less-organised layer of mesenchymal Thecal Cells.

Periodically a group of primordial follicles enter a stage of follicular growth. During this time, the oocyte undergoes a 500-fold increase in volume (corresponding to an increase in oocyte diameter for 10 μ in a primordial follicle to 80 u in a fully developed follicle). Concomitant with oocyte growth is an increase in the numbers of follicular granulosa cells, which form concentric layers about the oocyte.

Throughout this growth period, the oocyte remains arrested in the dictyate stage. The fully grown follicle thus contains a large oocyte surrounded by several layers of granulosa cells. Many of these cells will stay with the ovulated egg, forming the Cumulus, which surrounds the egg in the oviduct. In addition, during the growth of the follicle, an Antrum (cavity) forms, which becomes filled with a complex mixture of proteins, hormones, cyclic AMP, and other molecules. At any given time, a small group of follicles will be maturing. However, after progressing to a more mature stage, most oocytes and their follicles will die.

In order to survive, the follicle must find a source of gonadotropic hormones, and "catching the wave" at the right time, must ride it until it peaks. Thus, for oocyte maturation to occur, the follicle needs to be at a certain stage of development when the waves of gonadotropin arise. Day 1 of the menstrual cycle is considered to be the first day of "bleeding". This bleeding from the vagina represents the sloughing off of extra uterine tissue and blood vessels that would have aided the implantation of the blastocyst.

In the first part of the cycle (called the Proliferative or Secretory phase), the pituitary gland starts secreting increasingly large amounts of FSH. The group of maturing follicles, which have already undergone some development, respond to this hormone by further growth and cellular proliferation. FSH also induces the formation of LH receptors on the granulosa cells. Shortly after this period of initial follicle growth, the pituitary begins secreting LH. In response to LH, the meiotic arrest is broken.

The nuclear membranes of competent oocytes break down, and the chromosomes assemble to undergo the first meiotic division. One set of chromosomes is kept inside the oocyte and the other is given to the small polar body. Both are encased by the zona ellucida, which has been synthesized by the growing oocyte. It is in this stage that the egg will be ovulated.

The two gonadotropins, acting together, cause the follicle [is to produce increasing amounts of estrogen, which has'at tst five major activities in regulating the further progression the menstrual cycle.

1. It causes the uterine wall to begin its proliferation and to become enriched with blood vessels.
2. It causes the cervical mucus to thin, thereby permitting sperm to enter the inner portions of the reproductive tract.
3. It causes an increase in the number of FSH receptors in the follicular granulosa cells while causing the pituiary to lower its FSH production.
4. At low concentrations, it inhibits LH production, but at high concentrations, it stimulates it.
5. At very high concentrations and over long durations, estroge interacts with the hypothalamus, causing it to secrete gonadotropin releasing factor.

Thus, as estrogen levels increase as a result of follicular production, FSH levels decline. LH levels, however, continue to rise as more estrogen is secreted. As estrogens continue to be made (days 7 to 10), the granulosa cells continue to grow. Starting at day 10, estrogen secretion rises sharply. This is followed at midcycle by an enormous surge of LH and a smaller burst of FSH.

Experiments with female monkeys have shown that exposure of the hypothalamus to greater than 200 pg of estrogen per millilitre of blood for more than 50 hours results in the hypothalamic secretion of gonadotropin releasing factor. This factor causes the subsequent release of FSH and LH from the pituitary. Within 10-12 hours after the gonadotropin peak, the egg is ovulated.

Although the detailed mechanism of ovulation is not yet known, the physical expulsion of the mature oocyte from the follicle appears to be due to an LH-induced increase in Prostaglandin within the follicle. If ovarian prostaglandin synthesis is inhibited, the degenerative changes in the follicle leading to the release of the ovum do not occur, and no ovulation takes place.

Following ovulation, the Luteal Phase of the menstrual cycle begins. The remaining cells of the ruptured follicle under the continued influence of LH become the Corpus Luteum. (They are able to respond to this LH because the surge in FSH stimulated them to develop even more LH receptors.) The corpus luteum secretes some estrogen, but its predominant secretion is Progesterone.

This steroid hormone circulates to the uterus, where it completes the job of preparing the uterine tissue for blastocyst implantation, stimulating the growth of the uterine wall and its blood vessels. Progesterone also inhibits the production of FSH, thereby preventing the maturation of any more follicles and ova. (Such a combination of estrogen and progesterone has been used in birth control pills for this reason. The growth and maturation of new ova are prevented so long as FSH is inhibited.)

If the ovum is not fertilised, the corpus luteum degenerates, progesterone secretion ceases, and the uterine wall is sloughed off. With the decline in serum progesterone, the pituitary secretes FSH again, and the cycle is renewed. However, if fertilisation occurs, the trophoblast will secrete a new hormone, Luteotropin, which causes the corpus luteum to remain active and serum progesterone levels to remain high. Thus, the menstrual cycle enables the periodic maturation and ovulation of human ova and allows the uterus to periodically develop into an organ capable of nurturing a developing organism for 9 months.

EGG MEMBRANES

free kinds of egg membranes have been observed so far

(*a*) Primary egg membrane.

(*b*) Secondary egg membrane

(*c*) Tertiary egg membrane.

(a) Primary egg membrane

It is the innermost membrane the egg which is secreted by the egg itself during oogenesis. .s also called *vitelline membrane*. The vitelline membrane is rmed by the hardening of the outer peripheral layer of toplasm of the oocyte. Some time it is the only egg :mbrane present as in Hydra, earthworm and in mammals (placentals).

(b) Secondary egg membrane

More commonly outside the -imary vitelline, are present secondary and tertiary membranes. The secondary membranes are formed by the accessory follicular cells of the ovary closely associated with the egg. he membrane which is secreted by ovary is called the chorion in insect eggs and certain cephalopod eggs. In certain cases here it is tough, is provided with one or more than one aperture or micropyle for the entry of sperms.

(c) Tertiary egg menbrane

In the eggs of frogs, birds, ptiles are oviparous mammals, secondary egg membrane is enveloped by another hard or gelatinous material which constitutes the tertiary egg membrane. In frog it is made up of 4-6 gelatinous rings. In birds and reptiles, the white hard shell constitutes the tertiary egg membrane and is secreted by the inner wall of oviduct.

COMPARISON OF SPERMATOGENESIS AND OOGENESIS

From the above account it is quite clear that both the processes are alike to each other. The following compassion can be made between the two processes

Similarities

(*a*) Both the processes start from a primordial germ cell. (ii) Both pass through three phases, namely multiplication, growth and maturation.

(*b*) The process of synapsis or pairing of homologous phromosomes occurs in both the processes during the growth chase.

(*c*) In the maturation phase, both have two divisions.

Differences

(*a*) The spermatogenesis occurs in the testes while the oogenesis in the ovary cells.

(b) The process of spermatogensis results in the formation of four similar sperm cells from each primary spermatocyte. All these four are functional, while oogenesis results in the formation of only one large cell, from each primary oocyte, the egg cell and three non-functional small cells or polar bodies (polocytes).

(c) Mature sperm cells are very small and are also very active. On the other hand mature egg cell, has the large size and is very passive and receptive.

(d) All sperm cells are functional and due to much greater number of multiplication, the total number of sperms produced is enormously greater than the number of eggs.

(e) Complete process of spermatogenesis finishes in the tests while the oogoenesis is completed outside the ovary.

PHYSISO-CHEMICAL NATURE OF YOLK

Yolk is a complex of variably assembled components, rather than a definite chemical substance. The principal components are *proteins, phospholipids* and *fats* in different combinations. Depending on which of these components predominate, the yolk is distinguished as *"protien yolk"* or *"fatty yolk"*.

These two kinds of yolk are present side by side in the eggs of many animals. The avian yolk as a whole contains 48.7% water, 16.6% proteins, 32.6% phospbolipids and fats and 1 % carbohydrates. The fatty portion of avian yolk is mainly neutral fat (50% of the dry weight), the remaining being phosphatides and cholesterol. In animals, yolk is found in three forms as given hereunder:

Granular Yolk

Protein yolk of many invertebrates like echinoderms and of lower chordates *(Amphioxus,* Tunicates) consists of fine yolk granules which are fairly evenly distributed in the cytoplasm of the eggs.

Yolk Platelets

In *amphibians* eggs, the yolk is found in the form of large granules called *yolk platelets*. The yolk platelets are oval and flattened in one plane. They contain two main proteinaceous substances : *phosphovitin* and *lipovitellin.* Phosphovitin is a highly phosphorylated protein (phosphorus 8.4%), whereas lipovitellin contains a considerable amount of bound lipid (17.5%). In the yolk platelets 2 molecules of phosphovitin are associated with 1 molecule of lipovitellin in a strtuctural unit. Electron micrographs of amphibian yolk reveal that these units are arranged in the platelet in a crystalline lattice work with regular hexagonal packing.

In addition to the yolk platelets the amphibian egg contains lipid and glycogen. Lipid is found in the cytoplasm in the form of *lipochondria,* which consists of an internal core of lipid surrounded by a thin protein layer.

Cyclostomej, elasmobranchs, ganoids and the *lungfishes* have eggs with a distribution of food reserves much the same as in amphibians.

Yolk Spheres

The yolk of birds, reptiles and bony fishes lies in a compact mass in the interior of the egg. The cytoplasm is restricted to a thin layer on the surface, with a thickened cap on the upper side. Most of the yolk is liquid, but about 23% is in the form of solid *"yolk spheres"*.

FUNCTIONS OF YOLK

(*a*) Yolk is the most usual form of food storage in the egg.

(*b*) It influences the differentiation of ooplasm and the, patterns of cleavage.

(*c*) The size of the egg is determined by the amount of yolk present in it.

(*d*) Yolk exercises an important influence on the morpho genetic movements of blastomeres during gastrulation.

(*e*) The nature of development whether *indirect* with larval forms or *direct* with juvenile stages is governed by the amount of yolk present in the egg.

ROLE OF FOLLICLE/NURSE CELLS IN OOGENESIS

The growing oocytes are surrounded by special nutritive cells which immersely help their growth in various ways. There are two main types of nutritive cells, *viz., follicle cells* and *nurse cells.*

Follicle Cells

In mammals particularly and some other vertebrates, the oocytes are surrounded during growth and maturation phases by special cells of the ovary, the *follicle cells.* These are derived from the germinal epithelium of the ovary. Initially, the young oocyte is surrounded by a single layer of follicle cells but later, the number of follicle cells increases greatly, the cells becoming arranged in several rows.

In a mammal, the follicle cells and the developing oocyte together constitute a *grafian follicle.* As the egg approaches maturity, an eccentric cavity called *antrum* appears in the mass of the follicle cells This cavity is filled with a fluid known as *liquor folliculi* which is secreted presumably by the cells of the follicle.

In the beginning, there is a simple apposition of the follicle ells and the oocyte. The cytoplasmic membranes of the djoining cells remain separated by a narrow gap of about 80 Å. The plasma membrane of oocyte and that of follicle cells show lose connections at some points in the form of *desmosomes.* At a later stage a wider space appears between the follicle cells nd the oocyte.

However, the follicle cells maintain their ontact with the oocytes at the points where the desmosomes vere observed in the earlier stage. The surface of a young oocyte is drawn out into numerous finger like *microvilli* which roject into the space between the oocyte and follicle cells. these microvilli interdigitate with *cytoplasmic processes* of the ollicle cells. The presence of the microvilli greatly increases he surface area of the oocyte. This increase in area facilitates metabolic turnover between the oocyte and the surrounding ells.

It is believed that the follicle cells actively help in the growth of the oocyte by secreting substances which are taken ip by the oocyte. The individual microvilli cannot be seen with the optical ficroscope and the zone of microvilli appears as a radially striated layer known as the *zona radiata.* The presence of nail inpocketings of the oocyte cytoplasm at the base of the microvilli points out that the oocyte takes in fluid and dis)lved substances from the space between itself and the follicle ells by way of "cell drinking" or *pinocytosis.*

Nurse Cells

In some invertebrates like annelids, insects and molluscs the oocyte is surrounded, in addition to follicle cells, by special nurse cells. Derived from the egg cell, the nurse cells supplement the function of follicle cells in providing nutrition to the growing oocyte. In *Drosophila* an oogonial cell divides by four successive mitotic divisions into 16 cells.

One of these cells becomes an oocyte whereas the other 15 become nurse cells and nourish the oocyte. Nutrients from the cytoplasm of the nurse cells pass into the oocyte through gaps developed in the cell membranes two cell types. This type of relationship is different from that found between oocvte and follicle cells because in this case no microvilli or cytoplasmic processes are developed at the interface between the oocyte and the nurse cell. In some insects and annelids the nurse cells are gradually consumed during the growth of the oocyte. In molluscs e.g., the snail *Helix,* the entire nurse cells are engulfed in the cytoplasm of the oocyte.

ARCHITECTURE OF THE MATURE EGG

All eggs are surrounded by a cell membrane, or *plasmalemmma.* With very few exceptions, they bear as additional membrane, produced in the ovary, which is known as the *vitelline membrane;* in mammals it is known as the *zona pellucida.* Secondary membranes, secreted by ducts, such as the chorion of insects or the jelly coats of amphibian eggs, also occur but are not important to this discussion.

The cytoplasm of the egg contains, of course, various organelles, yolk granules, and often large amounts of pigment. Of considerable importance to the progress of development is a vaguely defined layer of gelated cytoplasm, approximately 2 to 3 microns in thickness, immediately below the cell membrane. This layer, including the cell membrane, is referred to as the *cortex,* or the *cortical layer,* of the egg. Its importance will be discussed later.

In most animals the cortex contains granules, of mucopolysaccharide nature, which range in size from less than 1 to approximately 2 microns. The cortex may also contain varying amounts of pigment. While the egg is still in contact with the follicular cells, follicular processes may penetrate the cortical layer ; this may be of importance in determining the specific character of the cortical layer in some animals.

Some organisation within the egg is indicated by its *polarity.* All animal eggs have a polar structure. The two opposite poles that can be distinguished are referred to as the *animal* and *vegetal poles* of the egg. An imaginary line connecting the poles defines the main axis of the egg, the *animal vegetal axis.*

The polarity of the egg is often apparent in the arrangement of its cytoplasmic inclusions. In many eggs the yolk is not evenly distributed; often its density increases from the animal toward the vegetal pole with corresponding structural ranges. In some cases there are substantial differences in the egree of pigmentation; in many amphibians, for example, are animal region of the egg surface is heavily pigmented, whereas the vegetal region is essentially unpigmented.

In addition to an obvious polarity, many animal eggs possess a *bilateral symmetry*. This may be expressed in the cape of the unfertilised egg, as in insects; in other cases the symmetry is not apparent until after fertilisation.

There is reason to believe that in many cases the polarity of eggs originates during their growth phase within the ovary. It is of some relevance, then, to inquire to what extent polarity of other evidence of organisation is expressed within the nfertilised egg and to what extent it is indicative of developtental potentialities or predestination.

Pigment, particularly in amphibians, is an easily recognised iarker of cytoplasmic asymmetry. Yet it is not clear that the figment distribution within amphibian eggs predelineates developmental fates. The pigment is localised most heavily at the very periphery cf the animal hemisphere of the egg, ithin the deepest layer of the cortex, and it does not extend ito the vegetal hemisphere.

There is no apparent corresponding differential of other cytoplasmic components of the frog egg; interestingly, however, during the developmental ages just prior to the deposition of yolk granules within the oocyte, the greatest concentration of ribonucleic acid coincides ith the localisation of heavy pigment. The distribution of igment in some mollusc and sea urchin eggs has some preictive value which will be discussed later. It may be worthhile to review briefly other evidences of organisation within ie unfertilised eggs of animals.

Selective localisation of cytoplasmic components, indicat ig some organisation of the egg, is best exemplified by tolluscs, annelids, and ascidians. In most molluscs the iperficial part of the egg cell is formed by a thin, yolk-free iyer of cytoplasm. Local thickenings of this layer may occur near the animal and vegetal poles, forming *pole plasms*. There are evidences of internal organisation also.

In certain gastropod molluscs, for example, movements of various egg substances occur during prefertilisation phases of maturation of the eggs, as a result of which they are concentrated at certain places within the ovum. These processes, termed *ooplasmic segregation,* establish a condition in which different parts of the egg differ more or less in their cytoplasmic composition.

Thus glycogen tends to shift and concentrate toward the animal pole of the egg ; ribonucleic acid also is of higher concentration in the pole plasm relative to the rest of the cytoplasm. Vitamin C, presumably bound to the Golgi bodies, tends to concentrate into an annular band near the equator of the egg. The displacements and selective accumulations of cytoplasmic components are continued during the early developmental stages, but it is noteworthy that these processes, important for later development, are well underway in the unfertilised egg.

There is little evidence for asymmetric distribution of cytoplasmic components within the eggs of higher animals. It may be noted that in the mammalian egg there is evidence for a heavy concentration of RNA at one pole of the egg. There is, however, no reason to believe that that pattern of pigment distribution and perhaps other suggestive distribution patterns represent a form of preorganisation in most animal eggs.

It is clear, nonetheless, that the animal egg is not to be regarded as a completely homogeneous system. Polarity and symmetry alone are difficult to explain without assuming local differences in the composition within the egg, which are apparent earlier in some species than in others.

ORGANISATION OF THE EGG

There are two extreme possibilities relative to the functional organisation of the egg. Either each organ in the adult is represented in the egg by some special physical or chemical composition, or the egg contains unorganised protoplasm that is gradually changed into the organised structures of the formed embryo.

The first concept *(preformation)* involves a mosaic, or a pattern of organ-forming substance, which are special substances that determine each and every organ. The second concept *(epigenesis)* depicts an unorganised protoplasm that requires the action and interaction of various agents to convert it into the various structures of the adult organism.

Before considering the additional implications of these two possibilities, it will be valuable to review conditions in the sea urchin egg. One local race of a mediterranean species was round to display a definite polarity indicated by a band of pigment composed of many small red granules. The parts of this egg essential to the following analysis are the thin outer cortex, the fluid, inner endoplasm, and the red band of pigment granules, the pigment band. The nucleus may be in any position in this cell.

The pigment band provides a natural marker for a specific region of the unfertilised egg; by observing this band during development, it is possible to determine what develops from the protoplasm of this specific region. The pigment band isjust below the equator of the egg, toward the vegetal pole. The first cleavage along the animal vegetal axis bisects the pigment band; during all the cleavage stages this band keeps its same position relative to the animal and vegetal poles.

At the later, eight-cell, stage, the pigment band is still visible. It has been cut off into the four lower cells in the vegetal hemisphere. It maintains its position just below the equator during the blastula stage, which consists of a single layer of cells arranged to form the walls of a liquid-filled sphere.

Figure shows a section of the blastula through the . pigment band and also outlines the fate of the pigment band through the process of gastrulation. For present purposes this process may be defined as a folding process that converts the single-layered blastula into a double-layered gastrula and forms the primitive gut, or *archenteron.* During the surroundings into the cytoplasm of the developing oocyte may account for its polarity.

In most cases the polarity of the egg is probably determined while it is still within the ovary. Whether it occurs through a kind of group effect, differential chemical gradients within the ovary itself, the transmembranal passage of materials into the oocyte, or a combination of these and other unrecognised mechanisms is subject to further investigation.

EGG MEMBRANES

The mature egg is covered by a cell membrane or *plasmalemma* which is the unit membrane. It is composed of an outer and an inner laryer of protein of 50 Å thickness. Between the proteinous layers, there occurs a lipidous layer of 60 Å thickness. In most animals, the plasma membrane is surrounded by certain special protective accessory egg membranes which are of following kinds:

Primary egg membranes

These membranes develop in the ovary between the oocytes and the follicle cells and may consist of mucopolysacchrides and fibrous proteins. Primary membranes are secreted either by the follicle cells or the oocyte may contribute in their for. mation in some cases

(*i*) In amphibians and birds the primary membrane is known as *vitelline membrane*. This membrane is actually of double origin. An inner layer of very rough fibres is produced in the ovary while the outer finely fibrous layer is formed when the egg enters the upper part of the fallopian tube. In tunicates and fishes it is usually called the *chorion.*

(*ii*) The primary envelope of shark, some bony fishes, some amphibians and some reptiles have striated appearance and is called *zona-radiata,* which represents the degraded microvilli of the growing oocyte. In fishes, perforations in the egg membrane left after the microvilli are withdrawn. These perforations in zona radiata become canals called the *micropyles* through which the spermatazoa can reach the egg. Zonaradiata is absent in bird's egg.

(*iii*) In mammals, the unstriated and modified zona-radiata is called *zona-pellucida*. It is formed by joint efforts of ova and follicle cells.

(*iv*) An egg while escaping from the mammalian ovary rries with it on the surface of zona-peilucida a layer of follicle cells known as *corona-radiata*. These cells of coronadiata are peeled off later as the egg descends the oviduct.

All these primary egg membranes usually adhere closely to ie surface of the oocyte. But at a later stage a fluid filled space may appear between the egg plasma membrane and the rimary egg membrane. This space is called the *peri-vitelline* space.

Secondary egg membranes

These membranes are secreted by oviducts and other sory parts of the genital organs when the egg passes from vary to the exterior.

(a) In oviparous sharks and rays the egg is surrounded by a *hard shell*, secreted by shell glands of the oviducts. The shell drawn out into long twisted horns which serve to entangle e eggs among sea weeds.

(b) The amphibian eggs are surrounded by a layer of *jelly* hich protects the egg and sometime serves to make the eggs to adhere to one another and to submerged objects such as plants. The *jelly* absorbs water and swells.

(*c*) Most complicated egg membranes are found in the eggs of birds, where no less than five membranes can be distinguished. In addition to inner most *vitelline membrane* which is primary in nature, the following are the secondary membranes:

The *white of egg* is the denser part of the egg. It forms two strands known as *chalazae* which help to keep the egg cell in the centre of the egg white.

Next are two layers of *shell membranes* consisting of fibres of keratin matted together.

The outer most membrane is the porous *shell* which consists chiefly of $CaCO_3$.

The egg membranes of a bird's egg are secreted one after another as the egg proceeds down the oviduct. These membranes not only serve for the protection of the egg cell but the egg white serves as an additional source of nourishment.

Significance of Envelopes of Egg

Regardless of their form, egg envelopes appear to be initially secretions of either the oocyte or the surrounding follicle cells, mainly as by-products of cell activity or as a basement membrane typical or epithelial tissue as such, each chemically peculiar to the tissue and species concerned.

Any material or structures once present may be exploited secondarily to serve other functions. Thus, material from either the oocyte or the surrounding follicle cells or some oviducal tissue, readily serves the eggs as protective structure and it is commonly augmented for this purpose, particularly in relatively large, yolky eggs and in all eggs that are laid out of water.

Even in mammalian egg-small, almost yolkless, and existing in a nutrient liquid medium when shed from the ovary-the *zona pellucida* surrounding the ovum, has been considered (1) as a protective device to prevent polyspermy, *i.e.* fertilisation by more than one spermatozoon ; (2) as a means of preventing egg fusion ; and (3) as a necessary means of maintaining normal cleavage 01 the egg following fertilisation.

Further, in ascidians, particularly solitary ascidians, not only does the basement membrane of the follicle cells survive as the chorion, which plays a role in preventing self-fertilisation of the egg, but the follicle cellh themselves persist attached to the chorion and, by virture 01 vacuoles in their cytoplasm, function collectively to increase buoyancy.

TYPES OF EGGS

Some eggs are small while others are larger. Eggs of certain birds are supposed to be the largest in size. Eggs are the modified and largest animal cells. The large size is due, not to a general exaggeration of all the parts normally found in an animal cell, but to the large quantity of cytoplasm that is present. In eggs small or good amount of yolk particles, synthesized by Golgi body, mitochondria or endoplasmic reticulum, are present. The *yolk* or *deutoplasm* serves as a source of nourishment for the young organism until it can fend for it self or alternatively exploit the food provided by the parent. On the basis of presence or absence and amount of yolk particles, the eggs can be classified as follows

Alecithal type

When the yolk particles are entirely lacking as in placentals.

Olligolecithal type

Here the yolk particles are present but in little amount as in Echinodermata.

Isolecithal type

When yolk particles are distributed evenly throughout the cytoplasm as in fishes.

Centrolecithal type

In insects the yolk particles are restricted to the centre of the egg.

Telolecithal type

When the yolk increases in quantity and becomes accumulated at one end or pole of the egg as in frog.

Macrolecithal type

In birds and repticles, the yolk particles are in good amount. Here, in such cases the size of nucleus and cytoplasm reduce and lie in the form of a small germinal disc.

The yolk present in the eggs of chordates is, as a rule, aggregated chiefly towards one pole of the cell, the *vegetative pole,* whilst the egg nucleus lies near the other pole, the *animal pole,* in a region relatively, or completely, free from yolky substance.

COMPARISON OF THE EGG AND THE SPERM

Both gametes are morphologically complete cells. Each has a nucleus and a cytosome containing representatives of the centrosomes, mitochondria, and Golgi bodies. Each has a plasma membrane. Yet neither is capable of independent, continued existence, for physiologically they are unbalanced.

The egg is larger, inert, contains a vast store of metaplasm, is protected by egg envelope,and has lost the power of continued division. The sperm is small, highly motile contains little cytoplasm and no metaplasm, is devoid of protective membranes, and in itself has lost thepower of continued division.

THE INFLUENCE OF YOLK ON EARLY DEVELOPMENT

As already stated that the eggs are supposed to be the largest cells in the body and are provided with little good amount of yolk or deutoplasm, which plays important part in determining the type of cleavage undergone by the fertilised egg. *Yolk* or *deutoplasm* is an inert material and when present in

quantity has a noticeable effect in damping down the rate at vhich the cleavage planes pass through the egg. If yolk was :venly distributed, then its presence would merely slow down livision, but its uneven distribution in telolecithal eggs causes he cleavage planes to be impeded chiefly in the vegetative kemisphere. This leads to the production of blastomeres of inequal size, those nearer the animal pole being smaller (beause they divide more rapidly) than those nearer the vegetative ole. In such cases the blastocoel becomes excentric in posiion and lies nearer the animal pole.

In fishes, reptiles and irds the large eggs are having great quantities of yolk and in every egg the yolk particles are separated from the cytoplasm which is confined to a small cap, the *blastodisc,* at the animal pole. Here in such cases only the cytoplasmic blastodisc being capable of cell division, whilst the yolky portion remains as such (undivided).

This type of cleavage in which the furrows do not pass completely through the egg, is spoken of as *partial* or *meroblastic* in contrast to the *total or holoblastic* cleavage experienced by eggs having a smaller concentration of yolk as in amphioxus, amphibians and most mammals.

5

Fertilisation

Fertilisation is the union of the male and female gametes and the formation of a single group of chromosomes. It is usually irreversible. This fusion establishes in the zygote the diploid condition characteristic of the species involved. The genetic sex of the individual is also determined at this time; the individual is male or female depending on whether the egg is fertilised by a sperm carrying a Y or an X chromosome.

When the sperm unites with the egg, the egg becomes activated and the zygote begins a series of cleavages that lead to the development of the embryo. The spermatozoon normally contributes the centrosome, which organizes the necessary machinery for cell division. In some cases the point at which the sperm enters the egg establishes the plane of bilateral symmetry in the future embryo.

In describing fertilisation, the sequence of events leading to the union of the two gametes is usually considered a part of the process and involves the following steps:

1. Activation of the spermatozoon, referred to as *capacitation*
2. Approach of the spermatozoon to the egg and its contact with the egg membranes
3. Penetration of the membranes surrounding the ovum
4. Activation of the egg
5. Syngamy or fusion of the pronuclei, and the establishment of the first cleavage spindle

Fertilisation has greater significance, however, than the union of the sperm with the egg and the production of a new individual. When it involves the union of chromosomes from unrelated groups, it also ensures the distribution of genes throughout the population. It is a method of gene dispersal that introduces variety, offering greater combinations for natural selection.

At this level of interpretation, fertilisation is a phenomenon that aids in adaptive variation, and is one of a series of nuclear changes that leads to the evolution of the species. Certain marine invertebrates, particularly the sea urchins, are the subjects of most of the studies of fertilisation. Their eggs are readily available, relatively large, lack obscuring yolk granules, and can be maintained with ease in their normal physiological medium, seawater.

Observation of the steps leading to the fusion of the egg and sperm is less difficult than in other forms. Among the vertebrates, fish and amphibian eggs provide good investigative material for analytical studies, but there are few extensive descriptions of fertilisation of mammalian eggs. Not only are these eggs difficult to obtain but there are also many problems in establishing and maintaining an environment suitable for the continued viability of the gametes. The factors involved in the successful fertilisation of the egg by the sperm have not all been defined.

In fact, the whole subject is somewhat controversial, with no general agreement as to the essential requirements or the steps involved in the fertilisation reaction. There is a tendency, therefore, to extrapolate and assume that the processes that occur in one form of animal life (marine invertebrates) take place in the mammals as well. This may not be true. The generalised picture of fertilisation presented in this chapter is a composite one and all events may not occur in all animals.

If we recognize the danger of overgeneralisation, the stepwise pattern of fertilisation described offers insight into an extremely complex process. Many features remain unknown, and many species need to be investigated, before a specific picture of fertilisation can be detailed for any one group of animals. The understanding of fertilisation is further complicated by the fact that although normally the penetration of the egg by the sperm triggers egg activation and the subsequent events leading to development, the same reaction can be induced by other means.

In nature, the eggs of a few animals, bees and wasps for example, usually develop without fertilisation. Activation of the egg, then, does not depend on some unique property of the spermatozoon, since a variety of agents are effective in initiating changes in the egg cytoplasm.

Such a phenomenon (that is, the development of an egg without fertilisation) is known as parthenogenesis, and when it occurs in nature it is referred to as *natural parthenogenesis*. When experimentally induced in the laboratory, it is *artificial parthenogenesis*.

For many years this phenomenon has intrigued biologists trying to gain insight into the problems of fertilisation by inducing development through artificial parthenogenesis. By this experimental technique they hope to separate those factors of development that are indigenous to the egg from those introduced by the sperm.

CONTACT OF EGG BY SPERM

Meeting of Gametes

For fertilisation to occur successfully, the male gamete first must find the female gamete; the ripe sperm must be delivered at the right time to the vicinity of the ripe egg. Fertilisation does not occur at the same stage in the life cycle of all unfertilised eggs. The eggs of various animals differ in their stage of nuclear maturation at the time of fertilisation. In some animals the germinal vesicle is still intact and sperm penetration initiates the meiotic process.

In others, the second polar body has formed and the female pronucleus is ready to fuse with the male pronucleus. The majority of vertebrate eggs are ovulated when they are in the second meiotic metaphase. As can be seen in Figure elsewhere in this chapter, the sperm can enter the egg anytime between the germinal vesicle stage (oocyte) and the stage when both polar bodies have formed (ootid).

The stage of nuclear maturation, however, at which a particular egg can be fertilised is specific for the animal involved. Several devices have evolved in vertebrates to ensure encounter between these two germ cells. The reproductive patterns of most animals involve the use of at least one such device. Light and environmental temperature affects the sexual maturation of many marine invertebrates, sea urchins for example.

Breeding cycles of vertebrates are under the control of hormones that direct the simultaneous maturation of the gametes and bring about their synchronous release. In many cases intricate patterns of mating behaviour that bring the adult female and males together at the proper biological time have evolved.

In terrestrial animals copulatory organs facilitate the delivery of sperm into the female reproductive tract near the site of egg release. The timing of gamete release is a critical one. Once ejaculated or

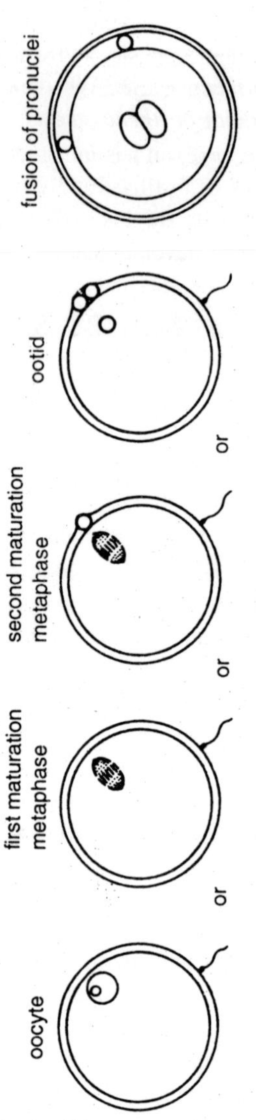

Figure 5.1: Stages of egg maturation and fertilisation. Sperm can enter the egg any time between oocyte stage and completion of meiosis. The stage of nuclear maturation at which a particular egg can be fertilised is specific for the animal involved.

ovulated, in general, the lifespan of the sperm or ova is short, ranging from a few minutes to a few days. The ova of many fish and amphibia, unless protected in some way, lose their ability to be fertilised within seconds or minutes after their release into the water. For example, those amphibia that deposit their sperm in packets (spermatophores) provide an environment in which the sperm remain viable outside the reproductive tract of the male for a long period of time.

Sperm released into pond water have a shorter functional life-span than do those shed in the more physiological saltwater. Gametes deposited in the female reproductive tract (internal fertilisation) remain fertilizable for an even longer period of time.

The ovulated human egg remains fertilisable for about 24 hours. The sperm also remains active for about that length of time. The fact that the sperm is so tiny, contains little or no energy stores, and expends energy as it moves about definitely limits its functional life. There are exceptions to this; for example, during hibernation, the spermatozoa of the bat may survive in the reproductive tract of the female for several months.

In the hen the sperm migrate the length of the oviduct where they collect in the ostium. A hen can lay fertile eggs for as long as 3 weeks after insemination, demonstrating that the life-span of spermatozoa in this species is longer than in many vertebrates. Man also manipulates the life-span of sperm by preserving them, usually under low pressure in sperm banks. This has been done for many years by livestock breeders who collect semen from prize studs for artificial insemination of breeding females.

Only recently has the idea of sperm banks for storage of human sperm been accepted, although the suggestion was made many years ago by Mueller who was concerned about the survival of the human population in an atomic age. Now the idea is correlated with the ban on population growth. Some males who plan to undergo vasectomy store their sperm in sperm banks. A reservoir is thus provided for their genetic material in case they change their minds about the number of children they wish to father. Eggs also can become overripe and unfertilisable.

They can remain in the uteri of the frog from 4 to 10 days and still be fertilised, but they lose this ability as the time increases. The complete degeneration of the eggs of most mammals follows ovulation without fertilisation. They apparently undergo fragmentation and are absorbed by the cells of the reproductive tract or are expelled through the vagina. Such factors as concentration of sperm, anaerobic or aerobic conditions, presence of glycolyzable substrates in the semen, and pH of the environment affect the life of the gametes.

For these reasons it is extremely difficult to find an artificial medium that simulates the normal environment of the reproductive tract, one in which the eggs can be fertilised and their developmental stages followed in vitro. Animals shed astronomical numbers of eggs and sperm if the gametes are merely released near one another into the surrounding water. It is estimated that the bullfrog, *Rana catesbeiana,* may lay as many as 20,000 eggs at a time. Even in more protected environments the male releases excessive numbers of spermatozoa compared to the number of eggs ovulated.

In a single ejaculation the human male may produce several million motile sperm, but only one sperm penetrates the plasma membrane of one egg. These large numbers of male gametes appear to be essential for successful fertilisation. Even the female reproductive tract is a hostile environment for the migrating sperm. Several barriers interfere with the progress of the sperm toward the *ampulla* of the *fallopian tubes,* where the union of gametes ultimately takes place. Cilia lining the tubes beat in the opposite direction, toward the uterus.

The lumen size progressively decreases from the uterus to the *ostium* of the fallopian tube, making it more difficult for sperm migration. The mucous secretions of the reproductive tract dilute the sperm; many are lost on the surface of the mucous lining. As a result, despite the excessive numbers produced, few sperm actually make their way to the site of fertilisation. Austin reports that he was able to recover only less than 100 sperm in the fallopian tubes of the rat, mouse, and hamster, about 200 in the field vole, and less than a thousand in the rabbit, dog, and sheep.

Although copulatory organs develop in terrestrial animals, the distance covered by the sperm before they reach the immediate environment of the egg is very great, considering the size of the gamete. The contractions of the muscles in the wall of the ductus deferens and urethra aid the deposition of sperm into the female genital tract of mammals. The human sperm travels from 3 to 6 mm. per minute. It takes 3 to 4 hours for human and rabbit sperm to reach their destinations.

In general, sperm move at a faster rate than can be accounted for by the simple movements of their flagella. Contractions of the muscular wall of the female reproductive tract propel the sperm forward. It has been suggested that the downward beating of the cilia in the tubes aids the sperm ascent by creating currents in the lumen of the tube. The contents of the tube flow down the sides and then circulate up through the center, presmnably moving the sperm along with it.

Apparently, spermatozoa use their flagella only when they have reached the vicinity of the egg. Random movements and chance dictate which sperm comes in contact with which egg. Since most sperm are motile, they must be deposited in an aquatic medium in order for them to reach the egg. Aquatic animals merely release their sperm in the water over the eggs at the time they are spawned, or soon after. The right environmental conditions stimulate large numbers of sea urchins to shed their eggs or sperm into the water at one time, forming clouds of gametes. Fertilisation takes place on a grand scale.

Insemination of frog eggs occurs as the female lays her eggs during *am plexus,* sexual embrace of the female by the male. The sperm must be in the immediate vicinity since the eggs are encased in jelly-like envelopes that absorb water. As soon as they reach the water, the jelly rapidly swells and is a handicap to further sperm penetration. As was pointed out earlier, some amphibians protect their sperm by packaging them in protective, moisture-conserving capsules that the female picks up by her *cloacal lips* and retains until needed.

Vertebrate animals, engaging in internal fertilisation, substitute semen for the water medium. It also supplies an energy source for the almost cytoplasmically naked spermatozoa. In many vertebrates the union of sperm and egg is further complicated by the various membranes that surround the egg. They must be pierced if the sperm is to make contact with the cytoplasmic surface.

In many invertebrates and a few vertebrates (such as teleosts and sturgeons), several small holes or micropyles in the membrane allow the sperm to pass through the thick outer covering of the egg, as seen in Figure elsewhere in this chpater. The lamprey egg has a small tuft of material at the animal pole and the chorion in this region lacks a thick jelly coat. It is at this point that sperm penetrate the chorion. In some manner the micropyles attract the sperm as shown by their increase in motility as they approach the egg. For the majority of the vertebrates, however, micropyles do not exist and the sperm must penetrate the jelly coats or other investing membranes.

Follicle cells still cling to the *zona pellucida* after ovulation of the mammalian egg and the minute spermatozoon must penetrate all layers.

Interacting substances of sperm and eggs

Most investigators agree that there is no definite evidence of chemotaxis in animals. Chemotaxis is usually defined as the ability of the sperm to detect a difference in concentration of some substance released into the water by the egg and to move from an area of lower to one of greater concentration of that substance.

Apparently, such a chemical attraction of the male gamete operates only in the plants, mosses, and ferns. There is a suggestion, however, that fish may have such a factor. The spermatozoa of these vertebrates become increasingly active as they approach the micropyle of the chorion,

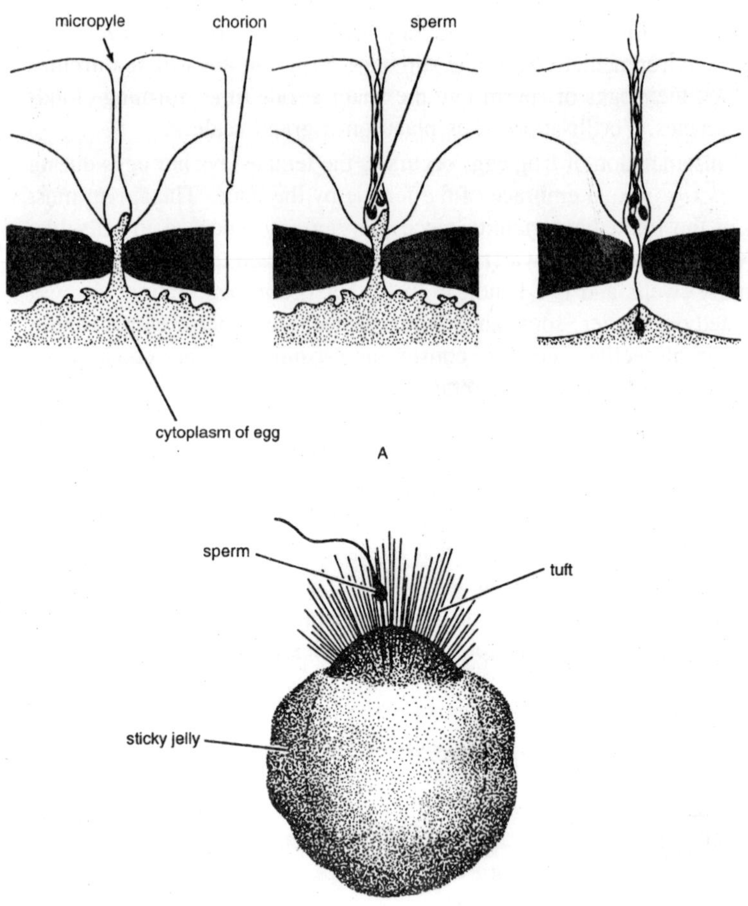

Figure 5.2: Thick membranes complicate the union of sperm and egg. A, Small holes, or micropyles, allow the sperm to pass through the outer covering of some teleost and sturgeon eggs. B, A sperm penetrates the chorion of the lamprey egg at region of tufts.

and they gather about its opening as if attracted specifically to its location. An isolated bit of chorion containing a micropyle also attracts sperm. On the other hand the vitelline membrane of the egg alone does not stimulate the sperm to greater activity or cause them to move toward the piece of tissue. The chorion, specifically the walls of the micropyle, appears to release some kind of substance that aids the spermatozoa in finding the opening leading to the egg. However, the substance has not been chemically identified.

We have known for many years that the eggs and sperm of some animals are chemically attracted to one another. They release into the surrounding medium substances that bring about agglutination of the sperm and the adherence of the sperm to the egg. These substances are considered to be more of a trapping agent for the sperm than a chemotaxic factor. In some cases they appear to act by preventing penetration of the egg by more than one sperm. Studies of these interacting substances have been carried out mostly on the sea urchin, but the eggs of vertebrates probably induce a similar agglutinating effect.

Such substances have been described for the cyclostomes, fish, and amphibians. F. R. Lillie, one of the pioneers in the field of fertilisation, studied the process in sea urchins and concluded that the jelly coats of the egg did more than protect. He found them to be saturated by an active material that he believed was released by the egg.

This material stimulated sperm motility and brought about a clumping of the sperm, usually head to head. If ripe unfertilised sea urchin eggs were allowed to remain in seawater for a time and then the eggs were removed and sperm added, the spermatozoa became sticky and adhered to one another. Lillie called the material located in the jelly coat *fertilizin.* A protein substance extractable from the surface of the spermatozoa was given the name *antifertilizin.* According to Lillie's fertilizin theory, presented in diagram form in Figure elsewhere in this chpater, fertilizin present in the cortex of the egg contains two active binding sites.

One is an ovophil group, the other a spermatophil. At fertilisation the receptor on the sperm surface combines with the spermatophil site. Lillie hypothesized that when this happens all the rest of the sperm-binding sites somehow become blocked and penetration of more than one sperm is thus prevented. The ovophil groups, in turn, react with the egg causing it to become activated.

Although all of Lillie's conclusions are no longer acceptable, his work stimulated many investigations in the field. A more modern interpretation of Lillie's interacting substances has been offered by Tyler. According to him, fertilizin is definitely located in the plasma membrane of eggs, extending into the egg's gelatinous coats (if present).

Antifertilizin can be detected in the membrane of the sperm. Both substances are specific receptor substances that interact in an antigen-antibody-like fashion. The reaction depends on the spatial arrangement of the atoms, in that one molecule fits the surface of another something like a key in a lock.

Tyler suggests a scheme for sperm penetration, shown in Figure elsewhere in this chapter, in which the interacting substances aid in

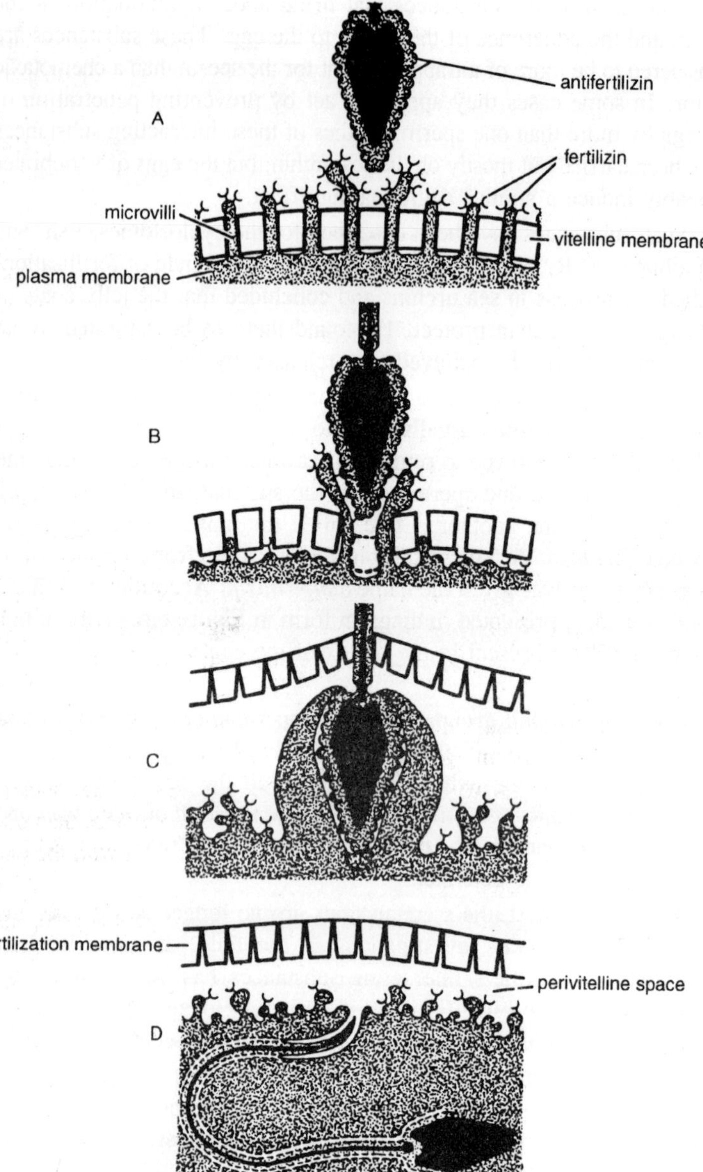

Figure 5.3: Scheme for sperm penetration.

the incorporation of the sperm into the egg cytoplasm. As detected in electron micrographs, many microvilli, extensions of the plasma membrane, perforate the vitelline membrane. Fertilizin located on the protruding tips of the microvilli reacts with the antifertilizin of the sperm.

After this initial contact with the fertilizing sperm, the microvilli retract drawing the sperm into the egg. Other sperm are unable to penetrate the vitelline membrane and enter the egg cytoplasm, since the receptors on the piasma membrane are inaccessible to them. Fertilizin is a glycoprotein that is probably synthesized in the gonad during meiosis, rather than by the egg as Lillie believed.

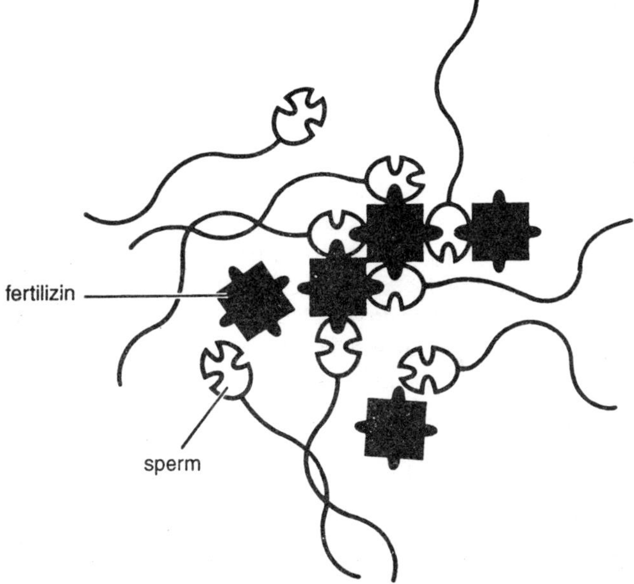

Figure 5.4: Suggested action of sea urchin fertilizin. Multivalent fertilizin can link several sperm together, producing clumping reaction.

It composes the jelly coat; it actually is the jelly coat and dissolves in the seawater after the eggs are laid. Since it is multivalent, it can link several spermatozoa together, resulting in the clumping reaction of sperm, possibly as shown in Figure elsewhere in this chapter.

Fertilizin consists of a number of amino acids and monosaccharides that vary from one species of animals to another and that are responsible for the species specificity of the material. The reactions between fertilizin and antifertilizin are stronger when they involve egg and sperm from the same species, although related species give weak reactions. The specificity of the reaction helps ensure the fertilisation of the egg by the same species of spermatozoon and not some other. One theory suggests that the fertilizin molecules on the egg surface trap the sperm and hold it at its surface, making penetration easier.

However, this is difficult to prove. All investigators do not agree that the fertilizin molecule is located on the egg surface as well as on jelly coats. Others believe that the reaction between the two molecules merely traps the sperm in the vicinity of the egg, inhibiting polyspermy and allowing one sperm to penetrate the egg surface.

Although most of the investigations have been carried out on invertebrates, there is evidence that the jelly coats of frog and toad eggs contain interacting substances that are both speciesspecific and shared (between species) and that they play a role in the penetration of the egg by the sperm.

The cells in the various regions of the frog's mviduct are responsible for secreting the jellies, with the upper third responsible for the jelly containing antigens shared by more than one species, and the lower levels of the oviduct producing species-specific antigens. The species-specific antigens then would be present in the outer jelly coat, since it is the last to be added. Shaver, in 1966, suggested a scheme of sperm penetration in which a reaction occurs between the species-specific molecules of the sperm and those in the outer jelly coat of the frog's egg, that is, the union of combining sites on the sperm surface with the jelly molecules.

This helps to ensure interbreeding by the same species. As a result the sperm is activated, or *capacitated,* a reaction described more fully later in this chapter. As the sperm increased its penetration distance through the jelly envelope, other combining sites on its surface would react with jelly molecules common to other species and the final reaction between complementary sites on egg and sperm would allow the last step in sperm penetration to occur.

There are some reports of the presence of interacting substances present in the eggs of rabbits, mice, and cows where the zona pellucida is evidently responsible for the production of the material. In general, interacting substances in the gametes of mammals have not been studied in any detail.

SPERM PENETRATION

The actual penetration of the egg by the spermatozoon can be separated into three general steps.

1. The sperm must penetrate the membranes that surround the egg.

These vary from the thick tough chorion or vitelline membrane of some fishes, to the jelly layers of the amphibian, and to the zona pellucida and attached corona radiata of the mammals.

2. When the spermatozoon contacts the egg surface, its membrane fuses with the plasma membrane of the egg.

3. The sperm nucleus along, with its middle piece is drawn into the cytoplasm of the egg. The tail may or may not be left outside the membrane or trapped in the perivitelline space.

The egg water of sea urchins not only brings about the agglutination of the spermatozoa, but it also causes the sperm to undergo a structural change known as the *acrosomal reaction.* Whether fertilizin is the responsible agent for all the changes associated with this reaction is not known.

The acrosomal reaction takes a variety of forms in different animals, but the end result is the release of materials, usually in the form of the acrosomal filament, and the secretion of lysins. The filament aids the sperm in penetrating the membrane barriers of the egg, and the lysins dissolve away parts of the membrane, providing a pathway for the entering spermatozoon. The solvent action of these enzymes provides the mechanism for sperm penetration.

Capacitation

Before the acrosomal reaction can take place, however, the sperm must undergo some kind of physiological change that increases its ability to penetrate the egg. The general term *capacitation* is given to this change. The egg water initiates this sperm activation in the sea urchin, and a similar process occurs in the amphibian when the spermatozoon comes in contact with the jelly coats.

Removal of the jelly coats from the egg before they come in contact with the sperm inhibits fertilisation. On the other hand, eggs taken from the body cavity, which lack jelly coats and normally cannot be fertilised, may be penetrated by the spermatozoon if they are first exposed to jellied eggs. Obviously the spermatozoa and the jelly coat somehow interact and increase the fertilizing capacity of the gamete.

The environment of the female genital tract induces capacitation in the sperm of those animals that practice internal fertilisation. In mammals freshly ejaculated sperm transferred to the ampulla region of newly ovulated eggs fertilizes few eggs. On the other hand, insertion of the sperm into the genital tract before ovulation occurs, activates the sperm in some way and enhances their capacity to fertilize the egg.

Acrosomal Reaction

Membrane fusion

Once the spermatozoa are activated, the series of events described as the acrosomal reaction may be followed by electron microscopy. Prior to the 1960s researchers believed that the plasma membrane of the egg elevated and phagocytised the sperm, drawing it inside the egg. Although this may occur in some instances, electron microscopic studies reveal that in general the penetration of the egg by the sperm involves a fusion of their membranes.

In this process the shape of the acrosome is altered. The successive stages in the acrosomal reaction in the invertebrate *Hydroides hexagonus* is depicted in Figure elsewhere in this chapter. The acrosomal vesicle (the covering over the acrosome) splits, the acrosomal granule disappears (apparently releasing lysin), and the one or more filaments contact the plasma membrane.

Invagination occurs in the plasma membrane at the points of contact with the filaments, and fusion of the two membranes takes place at these areas of interdigitation. These and other detailed studies of the acrosomal reaction have been carried out by the Colwins, who investigated this process in the annelid worm *Hydroides* and in the protochordate *Saccoglossus*. According to the Colwins, "the major role of the acrosome is apparently to deliver the sperm plasma membrane to the egg plasma membrane".

A similar series of acrosomal events have been observed in the toad, chicken, hamster, rabbit, guinea pig, and man; so it appears that the *Hydroides-Saccoglossus* pattern of gamete membrane fusion described by the Colwins also occurs in the chordates. Few complete studies, however, have been made on the vertebrates.

Mammalian spermatozoa recovered after they have penetrated the zona pellucida lack an acrosome. This loss is usually interpreted as analogous to the splitting of the acrosomal vesicle when the invertebrate sperm contacts the egg. The acrosomal granules released apparently aid the sperm in penetrating the thick pelluoida. One of the most detailed studies of sperm penetration in the mammal was made by Szollosi and Ris (1961).

They followed later stages in the fertilisation process in the rat and described the penetration of the plasma membrane by the sperm of this animal. Their electron micrographs showed that after the sperm penetrated the ovum it had no plasma membrane, although one was present before and during the process of fertilisation. They suggested

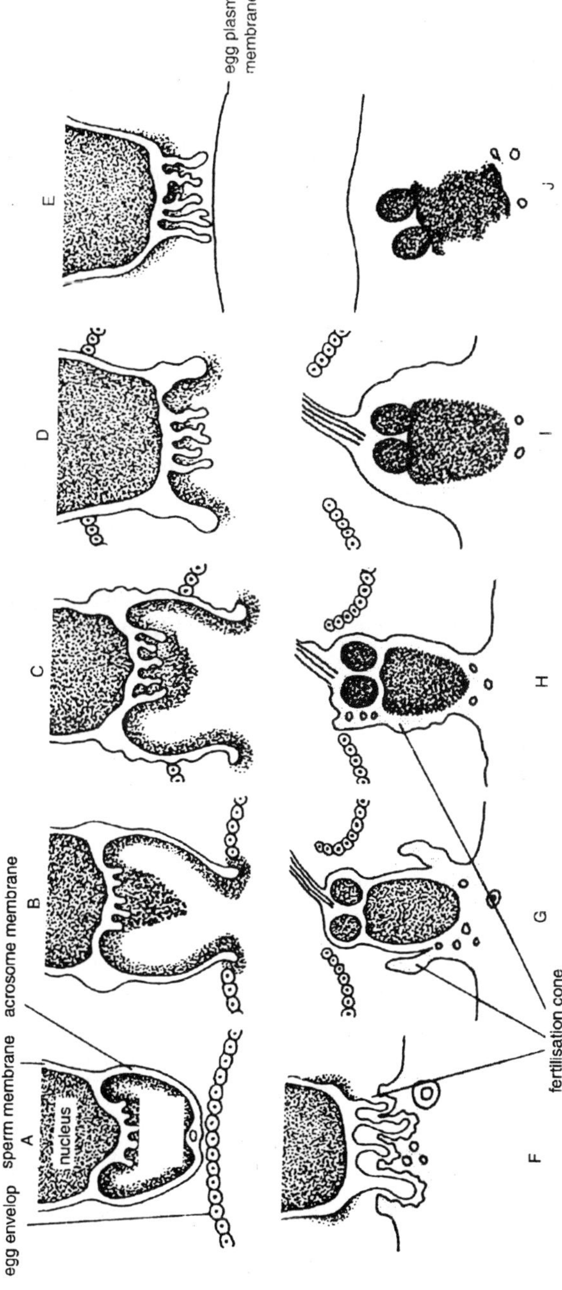

Figure 5.5: Sperm penetration in the marine annelid Hydroides. The acrosome membrane of the unactivated acrosome.

that contact between the sperm head and the plasma membrane of the egg brought about a rupture of the two membranes. As a result, the egg membrane fused with the sperm plasma membrane at the point of contact and formed a continuous membrane around the egg and outer surface of the sperm.

At first a deep fold of membrane forms around the sperm. Eventually, as fusion of membranes proceeds toward the tail of the sperm, the entire naked sperm is drawn into the egg, leaving its plasma membrane to be incorporated into the egg brave. These investigations by the Colwins and by Szollosi and Ris demonstrate the importance of egg and sperm plasma membranes in the actual entrance of the sperm into the egg cytoplasm. It is possible that the acrosome of both invertebrate and vertebrate sperm undergoes similar changes.

However, at the present time insufficient evidence on enough species prevents us from making a generalised statement concerning acrosomal structure, sperm activation, and egg membrane penetration. Once the acrosomal filaments contact the plasma membrane, the cytoplasm flows up inside the filament membrane, covering the sperm head, midpiece, and possibly the tail. This cytoplasmic elevation is the *fertilisation cone*. The type of acrosomal filament varies from animal to animal, ranging from one in *Arbacia* to many in the annelid *Hydroides hexagonus*. The shape of the fertilisation cone differs according to the form of the acrosome filaments.

In mammals, it has been described as a small mound at the site of sperm contact. The fertilisation cone may be of short duration, lasting only 20 seconds. In some invertebrate eggs the cone forms before fertilisation; thus doubt is cast on its function. In other examples the sperm appears to penetrate the membrane by undulating movements of the tail and no cytoplasmic cone forms at all.

Release of Lysins

The spermatozoa of most vertebrates contain a substance that enables them to penetrate the various layers around the egg. The material dissolves a localised pathway through the membranes enabling the acrosomal filament to reach the egg surface. The Colwins have depicted this lvtic activity in figure elsewhere in this chapter, B. This substance is enzymic in nature and known by the general name of *sperm lysins*. Lysins differ from one species to another and are released when the acrosomal granule disappears.

Attempts to extract the lysins and identify them chemically have not been too successful because of the small amount of material available.

They are protein in nature according to chemical analysis, but the specific class of enzymes has not been identified. Some investigators feel that the sperm may release more than one kind of enzyme.

According to recent studies, the jelly layers of certain amphibian eggs depend on the presence of disulfide bonds for maintaining their structure. It has been suggested that the lysins of this group of animals may utilize a disulfide bond as a substrate. According to these workers, the fusion of the egg plasma membrane and sperm membrane occur by a sulfhydryl-disulfide bond interchange.

They believe that fertilisation should be explained on a molecular basis. The mammalian egg offers a unique problem of sperm penetration, since at ovulation it is surrounded by the zona pellucida, the corona radiata, and various numbers of cumulus cells. It is very difficult to pull away the cumulus oophorus from a newly ovulated rodent egg although the sperm must penetrate this barrier. Its entrance is made easier by a lysin, a cumulusdispersing factor, identified as hyaluronidase, present in living rabbit, rat, and mouse sperm suspensions or extracts.

This dispersing factor acts upon the intercellular cement of the follicular cells by depolymerizing. and hydrolysing the *hyaluronic acid* cement that holds the cells together. In vitro studies show that the concentration of the enzyme is proportional to the number of sperm present, since it is associated with the sperm itself rather than the semen.

After some hours in the female reproductive tract the spermatozoa undergo capacitation and release their enzymes into the seminal fluid.

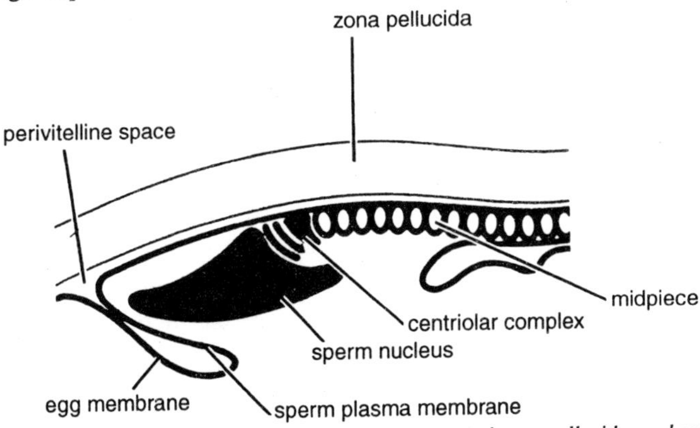

Figure 5.6: Sperm penetration in rat. Sperm has penetrated zona pellucida, and sperm plasma membrane can be seen to be continuous with egg plasma membrane. Naked sperm is drawn into egg.

Researchers assumed that the concentration of the enzyme was high to be effective in denuding the ovum, but they later learned that the sperm of the dog contains no hyaluronidase. This enzyme, then, is not the total answer to sperm penetration. It is now believed that the sperm probably contains sufficient enzyme to make a pathway for itself through the various layers.

Narrow slits in the zona pellucida appear after the sperm penetrates this membrane and offer visual evidence of a lytic substance. Microvilli anchor the cells of the corona radiata to the zona pellucida of the ovum. The two cannot be mechanically separated without injury to one of the membranes. The sperm may release more than one enzyme to act on the two layers.

The cells fall apart after contact with sperm or contents of the oviduct for a few hours. Some investigators suggest the possibility that a tubular factor that acts on the cells of the corona radiata is also secreted. Apparently after contact with the contents of the fallopian tube, the microvilli are retracted and free the ovum. Possibly the ciliary and muscular movements of the oviduct may mechanically remove the coronal cells as the egg makes its way down the tubes to the uterus.

EGG ACTIVATION

The sperm always enters the animal hemisphere of the amphibian egg, but in amphioxus penetration occurs at the vegetal pole. In other animals there are no restrictions. Sperm may enter the human ovum, for example, at any site on its surface. Once the sperm penetrates the egg membranes and deposits the compact nucleus inside the egg by the acrosomal tube, the male nucleus undergoes several changes.

The compact nucleus becomes hydrated and forms the male pronucleus, which is larger than the female pronucleus in several forms studied. The head of the sperm usually rotates 180 degrees so that the centrosome of the midpiece leads the way when the male pronucleus migrates toward the female pronucleus.

The centrosomes duplicate as the sperm head becomes hydrated and astral rays develop in the egg cytoplasm around the sperm aster. It is the sperm centriole that must organise the mitotic spindle if the egg is to cleave in its normal manner. In those eggs in which the maturation process was completed by the time the sperm penetrates the egg, the sperm immediately moves to meet the female pronucleus.

However, in those in which the completion of meiosis is a part of the activation process, the sperm remains quiescent, once it enters the ooplasm, until the second polar body forms. Evidently some change occurs in the cytoplasm (reflecting the stage of nuclear maturation) that signals hydration of the sperm and the appearance of sperm asters. At the time the eggs are shed, regardless of the stage of nuclear maturation, they usually become quiescent and enter developmental arrest. There is no new synthesis of RNA, and no other developmental changes occur unless the egg is penetrated by a sperm.

The arrest appears to be of a chemical nature since no new proteins are synthesized. The release of this arrest by the entrance of a sperm or by artificial means is referred to as egg *activation.* We do not know whether activation is caused by mechanical or chemical stimulation by the sperm. Once activated, however, the egg embarks on a series of changes that leads to the development of an embryo.

These changes involve the cortex as well as the more internal regions of the egg and are both structural and functional in nature. The immediate effect of sperm entrance is the completion of meiosis, if this has not yet occurred. The end result of egg activation is the fusion of the male and female pronuclei and the reestablishment of the diploid state characteristic of the species involved.

Completion of Meiosis

Usually the formation of the second polar body must be completed, since in most vertebrates arrest of the meiotic process occurs while the eggs are still in the ovarian follicle. Ovulation usually begins after the spindle forms for the second meiotic division, and the dyads arrange themselves on the metaphase plate.

The penetration of the sperm stimulates the chromosomes on the maturation spindle to pass into the anaphase stage. The completion of nuclear maturation immediately follows the fertilisation process. Little information is available concerning the details of the extrusion of the second polar body. We do know, however, that in the rabbit, as long as 45 minutes may elapse between sperm penetration and completion of the second polar body. It may take 2 hours for the mouse, and 4 hours for the rat to complete the meiotic process.

Cortical Reaction

The penetration of the egg by the sperm produces an almost instantaneous reaction in the cortex of the egg. These changes start at the point of sperm contact and pass over the surface of the egg; the exact changes that occur vary from animal to animal. In the past most of our information was based on microscopic examinations with dark-field illumination or polarised light. Studies show that the interference colour changes at the time of fertilisation, indicating a dispersion of material at the egg surface, possibly lipids. Often there is streaming of *cytoplasm* toward the point of sperm entrance and rearrangement of cytoplasmic constituents within the egg. Cortical granules visible at the surface of the egg disappear, and a membrane, the *fertilisation membrane,* lifts off the egg surface. Not until the electron microscope became available was it possible to obtain more precise information.

These new studies reveal that the surface of the sea urchin egg consists of two membranes, an outer vitelline membrane and an inner plasma membrane. At fertilisation the inner membrane remains on the egg and a space, the *perivitelline space,* develops between the two. Granules enclosed in membranes and closely associated with the egg membrane are located in the cortex of the egg.

These cortical granules explode and disappear at the time of fertilisation. Electron micrographs reveal that both the granule membrane and the egg plasma membrane rupture at the point of contact and the two membranes then become continuous as shown in Figure elsewhere in this chapter.

As a result the granules are ejected into the newly established perivitelline space where they line up and become attached to the vitelline membrane. The vitelline membrane and the material of the cortical granules contribute to the fertilisation membrane in these animals.

Cortical granules or vacuoles that disappear after fertilisation have been reported in the eggs of teleostean fishes, amphibians, hamsters, and rabbits. The granules of the fishes and amphibians contain a polysaccharide, similar to that identified in the sea urchin egg.

A lack of sufficient information prevents us from defining their specific role in the fertilisation process. However, the inference can certainly be drawn that the activity of the granules at the time of fertilisation is an integral part of egg activation. Fertilisation membranes similar to those described for the sea urchin have been studied in the protherian mammals, the frog, and the trout.

The vitelline membrane contributes to the fertilisation membrane in the frog; a similar membrane, the chorion, helps form the fertilisation membrane in the trout. Instead of a vitelline membrane or chorion, it is the zona pellucida that surrounds the eggs of higher mammals. There is little evidence that the mammalian membrane actually elevates at fertilisation although the egg shrinks and a perivitelline space is established.

However, it does undergo certain structural changes that can be correlated with entrance of the first sperm and that make it less permeable to other spermatozoa. After fertilisation, the egg's surface becomes smooth, and the microvilli of the corona radiata are no longer visible. Regardless of the specific type of outer membrane (vitelline, chorion, or zona pellucida), the surface of the fertilised egg and the unfertilised egg differ structurally from one another. We do not understand the mechanisms involved in the formation of the fertilisation membrane. It has been suggested that the "exploding" granules release substances into the perivitelline space to cause an influx of water, an osmotic phenomenon that might explain the growth of this space.

As the material leaves the cortex of the egg, some shrinkage in ooplasm occurs and this, too, could contribute to the formation of the space. The lifting of the vitelline membrane from the egg surface and the explosion of the cortical granules are energyrequiring reactions and thus can be prevented by inhibitors of oxidative phosphorylation.

Polyspermy Block

When a sperm penetrates an egg, certain physiological changes occur that can be correlated with the structural changes at the egg's surface. Once the sperm enters the egg, the egg surface becomes refractory to other sperm. The condition whereby the egg is penetrated by more than one sperm is known as *polyspermy* and is usually lethal to the cleaving egg.

If two sperm with their centrioles enter the egg, a triaster or tetraster forms, which directs the formation of two or more spindles. The chromosomes become distributed unequally on the spindles, and cell division soon comes to a halt. Therefore, after the penetration of one sperm a barrier, which restricts the entrance of another sperm, usually develops.

In the sea urchin this polyspermy-preventing barrier appears to reside in the fertilisation membrane, since its mechanical removal allows multiple sperm entry. In other animals the egg surface acts as the barrier, but in many mammals it is the zona pellucida that becomes impermeable to more than one sperm.

This reaction is known as the *zona reaction*. If polyspermy does result, the sperm usually remain in the perivitelline space. The presence of the sperm in the space seems to indicate that the plasma membrane acts as a secondary barrier. In a few cases polyspermy is normal.

More than one sperm enter the eggs of certain insects, amphibians, reptiles, and birds. In these animals, however, only one sperm develops into the male pronucleus and, therefore, only one cleavage spindle forms.

The precise polyspermy-blocking reaction is not understood. Since it is correlated with changes in the cortical region of the egg, the molecular rearrangement occurring at the surface of the egg at the time of fertilisation may be responsible in some way in preventing entrance of more than one sperm.

The perivitelline fluid that is formed, partially at least, from the discharged cortical vacuoles, appears to be effective also in blocking the entrance of more than one sperm. Vacquier et al. (1973) recently demonstrated the presence of a protease in the cortical granules of sea urchin eggs.

The protease hardens the vitelline membrane and lytically releases supernumerary sperm bound to the egg before the cortical reaction. The protease thus prevents polyspermy by this dual action.

Parthenogenesis

Activation of the egg does not depend on some unique property of the spermatozoon, since a number of agents are also effective in initiating changes in the egg cytoplasm. The fact that an egg can be stimulated artificially indicates that it contains all the essentials for the developing embryo.

In a number of organisms, eggs normally develop without fertilisation, by natural or spontaneous parthenogenesis. Aphids, wasps, and the male honeybee are examples. Most eggs, however, can be stimulated in the laboratory to undergo cleavage or *artificial parthenogenesis,* but only a few of these eggs complete their development. A large variety of substances can trigger the activation mechanism.

The eggs will react to chemical treatments, temperature shocks, pH, puncture with a fine needle that introduces foreign proteins, vibrations, and ultraviolet radiation. This long list of agents capable of inducing development indicates that no one common feature brings about the stimulus, but that the controlling factor rests in the egg cytoplasm itself. The initial response of these artificial stimuli is the cortical reaction, and it can be obtained by many of the stimuli.

This first response is followed by a second reaction, from the nucleus. Cleavage is more difficult to obtain. If cleavage follows the stimulation, some kind of substitute must be made for the centriole since it is usually contributed by the sperm and is responsible for formation of the cleavage spindle.

In some cases the spindle that has formed for the second meiotic division becomes more centrally located and serves as a cleavage spindle. In other cases the spindle appears to form spontaneously. One would expect that animals that result from artificial parthenogenesis would be haploid.

Although this is true for many of them, the diploid state is established in several of the more viable forms. The egg employs a variety of techniques to produce diploidy. One method involves a lack of cytoplasmic cleavage after the chromosomes have duplicated. A new nuclear membrane forms around the duplicated set of chromosomes and thus gives rise to an egg with a diploid nucleus. In some cases the fusion of female pronuclei with the polar body establishes the diploid state.

6

SPECIALISED DIVISION IN DEVELOPMENT

Biologists have demonstrated again and again the uniformity of living things. The morphologists point out common building plans; the biochemists describe similar metabolic cycles; the geneticists unravel common pathways in the transcription and translation of genetic information. The embryologists also show uniformity in the developmental processes of animals. The maturation of the gametes in the gonads as well as those processes leading to the establishment of a basic embryonic form, follow similar developmental pathways in all vertebrate animals.

The resulting body plan of one animal resembles that of another. The primordia (rudiments) of future organs develop and are arranged in a similar pattern. During evolution of the vertebrates, the hereditary factors controlling the generalised pattern of development have been retained. After the establishment of general structures, more specific ones that express the diversity of the species follow. Development always proceeds from the general to the specific.

The formation of *a blastula by* all vertebrates is an example of this generalised type of development. Although the specific size and shape may vary, all vertebrates form a blastula as the next stage after fertilisation. The development of a normal, adult animal from a zygote involves a series of distinct steps in which specific tasks are accomplished.

Cleavage, the mitotic division of the egg, is the first step after fertilisation and is a universally occurring phenomenon among metazoan animals. A series of rapid cell divisions enables a fertilised egg to develop into a complex multicellular animal, perfect in every detail.

Penetration of the plasma membrane by the sperm normally activates the egg and stimulates it to start dividing.

Fertilisation removes the block to cellular division that existed in the egg prior to fertilisation, and initiates a type of mitosis. As a result, cleavage brings about:

1. Partition of the cytoplasm into smaller units, *blastomeres*
2. Replication and distribution of nuclear genes throughout the cytoplasmic substances
3. A *blastula stage* consisting of groups of geographically segregated cells surrounding a cavity, the *blastocoel*

MECHANICS OF CLEAVAGE

As early as the late nineteenth century Boveri suggested that the sperm centriole controlled cleavage and that fertilisation introduced the centriole into the egg. Under the light microscope the centrioles appear as granules or rods located in the center of a clear area of cytoplasm, the centrosome. In electron micrographs, however, instead of single granules, they are shown to exist in pairs, with the long axis of one centriole of each pair perpendicular to the other.

At present there is no explanation for the right-angle arrangement. The centrioles are hollow cylinders, 1500 Å in diameter and 3000 to 5000 Å in length. Nine sets of triple tubes, extending the length of the cylinder, compose their walls. There is no evidence that the centrioles arise by division, or budding, of preexisting ones, or that they have genetic continuity as stated in the older literature.

Instead, they appear to be self-duplicating bodies that form anew from precursor molecules in the surrounding cytoplasm. However, the daughter centrioles always assemble in close relationship with the mother centriole. During the interphase, or possibly as early as the end of the preceding telophase, the centrioles replicate; they then take up their position on each side of the nucleus, where radiating, tubular, astral rays develop about them.

Each cell at metaphase, therefore, contains four centrioles, two at each pole, with each member of the pair oriented at right angles to the other. Embryologists generally agree that the centrioles organize the asters and mitotic spindle of the fertilised egg. As the centrioles migrate to the opposite side of the nucleus, a fiberlike connection appears between the two centriolesthe beginning of the spindle.

The synthesis of the microtubules composing the spindle fibers of the mitotic apparatus is the essential activity of sperm centrioles. These

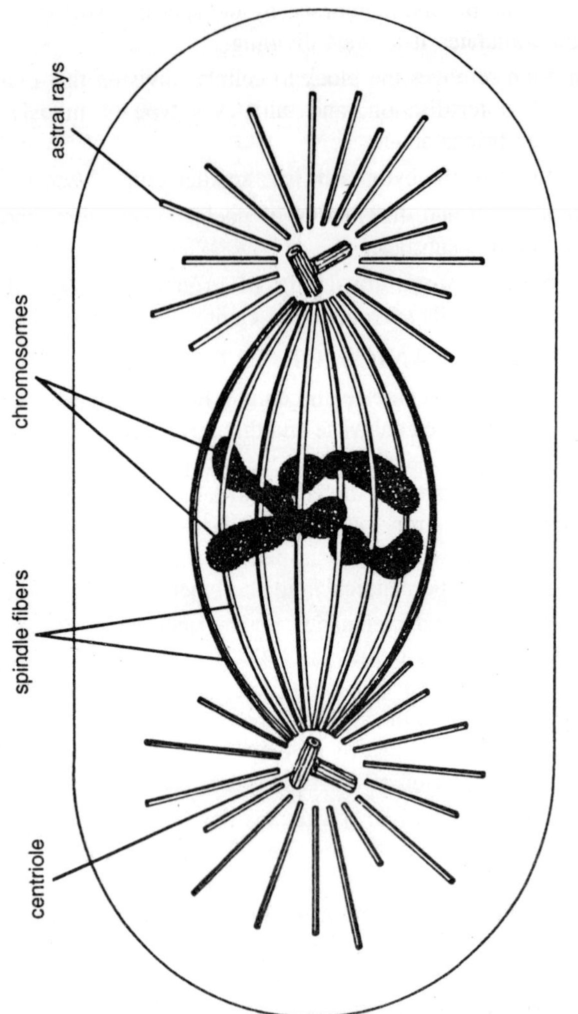

Figure 6.1 : Schematic representation of cell at metaphase showing two centrioles at each pole, with each member of the pair oriented at right angles to the other. Some spindle fibers pass from pole to pole, others from pole to centromere of chromosome.

fibers, or microtubules, extend from pole to pole. Electron micrographs also confirm the attachment of spindle filaments to the centromeres (or kinetochores) of the chromosomes.

At the end of the telophase most of these tubules disappear. Early cytologists suspected that the formation of the cleavage furrows and the division of the cell were under the control of chromosomes, since their movements were closely coordinated with the events of cell division.

However, it now appears that the mitotic apparatus controls both the movement of chromosomes and the formation of the cleavage furrow.

Surgical removal of the chromosomes does not inhibit blastomere division. So, obviously, the chromosomes themselves do not control furrow formation. When the cell is about to divide, the cortex becomes progressively stiffer, and furrows tend to form between each pair of centrioles.

These are especially evident when the egg cleaves unequally, for example, when the polar bodies split off from the egg. The furrow forms between the two centers even though the centrioles and their spindles are near the surface of the egg. As a result of this eccentric location, the cytoplasm divides into a large cell and a small polar body.

It is possible with modern techniques to move the metaphase spindle around inside the egg. Movement of the spindle more centrally results in a much larger polar body. By moving the mitotic apparatus about, a furrow can be induced to form at any point on the egg surface. Although the centrioles induce the furrow to form at a specific location, we do not understand the exact mechanism by which this is accomplished. It may vary with different species.

One theory suggests that the spindle and the cell surface interact, inducing the cortex to become activated, that is, to actually grow and form a contractile ring in the region of the furrow. The spindle can be removed once the ring is established, provided that mid-anaphase has been reached, and cleavage will continue according to the original orientation of the spindle. Removal of the spindle before mid-anaphase, in the sea urchin egg, inhibits division.

The relationship between spindle and cortex may be an actual physical one. The astral rays in the *Crepidula* egg (a mollusk), for example, are physically attached to the plasma membrane. Some evidence suggests that the influence exerted by the spindle may be attributable to diffusion of some substances from it to the cortex of the cell.

PARTITIONING OF CYTOPLASM

The fertilised egg rapidly divides into smaller and smaller units or cells, usually referred to as *blastomeres*. Cleavage produces hundreds or thousands of blastomeres. The rapidity with which one cell division follows another is a major difference between this type of cell division and ordinary mitosis. The mitotic cycles of cleavage have a much shortened interphase. During the period of cell division, there is little or no increase in total cytoplasmic mass.

The cells actually decrease in volume with each division until they reach the size of the normal adult cell characteristic of the particular species. This lack of growth after cellular division is a second difference between cleavage and ordinary mitosis. In normal mitosis, after the cells divide, a period of growth occurs. There is replication of chromosomes as well as synthesis of cytoplasmic materials.

As a result, the newly formed cells regain the size of the original mother cell. In cleavage, on the other hand, nuclear synthesis goes on, but little or no cytoplasmic growth occurs. For example, the blastula, consisting of thousands of cells, is usually only slightly larger than the unfertilised egg.

Thus, cleavage can be looked upon in part as a mechanism whereby the cell returns to the normal size for a particular species. The fact that no cytoplasmic growth occurs does not mean that the cytoplasm is synthetically inactive. Fertilisation (which marks the zero time of embryonic differentiation) releases the inhibition of the egg, probably by unmasking messenger RNA stored in the cytoplasm.

There is evidence that during maturation, the oocyte manfactures and stores in its cytoplasm information on how to synthesize the proteins required by the cell during cleavage. These proteins are needed for a variety of reasons. Cell membranes must be manufactured to cover the newly exposed cell surfaces formed by cleavage.

It is estimated that the first division increases the surface area by as much as 28% over that of the uncleaved egg. New proteins are also needed to construct the microtubules of the spindles. Further, some proteins, nuclear proteins for example, are apparently incorporated into the nucleus as histones (basic proteins linked to DNA) and RNA polymerase (responsible for mediating the transcription of genetic information from DNA to RNA). The removal of the zygote nucleus from sea urchin eggs or the addition of actinomycin D (which inhibits the synthesis of RNA) to egg suspensions, does not prevent fertilised (or artificially stimulated) eggs from synthesizing new proteins through the blastula stage (Denny and Tyler, 1964; Gross, Malkin, and Moyer, 1964). These experiments strongly suggest that the manufacture of new RNA does not occur during this period. Instead, stable RNA molecules stored in the cytoplasm during the maturation process carry the embryo through the cleavage stages.

About the time the embryo reaches the blastula or gastrula phase of development, the diploid nucleus directs the synthesis of messenger RNA (mRNA), and it is this new mRNA that will lead the embryo through differentiation. Two kinds of proteins, therefore, are synthesized by the cytoplasm, and their production follows a time sequence. The first proteins relate to cleavage, whereas the later ones involve embryonic

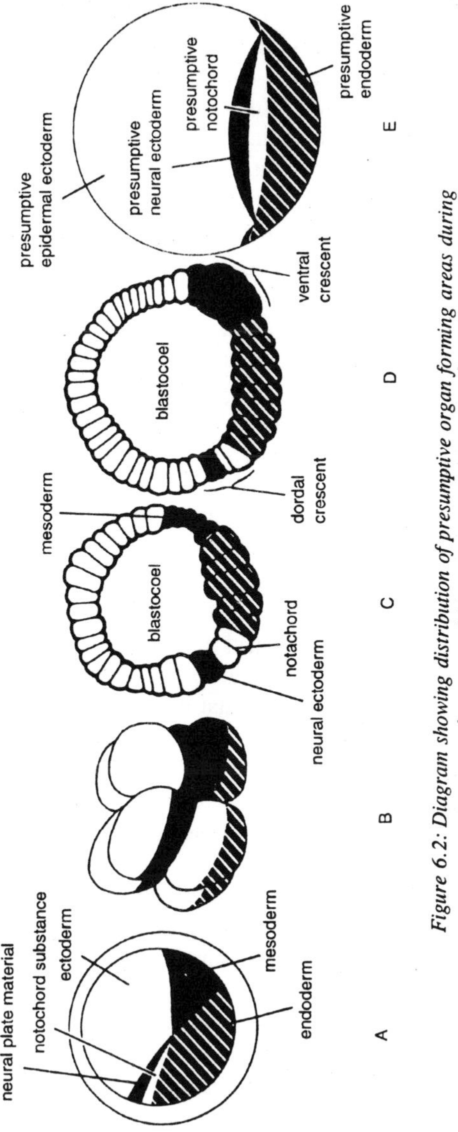

Figure 6.2: Diagram showing distribution of presumptive organ forming areas during cleavage and blastulation in amphioxus.

differentiation. As indicated in other chapter of this book, during cytoplasmic maturation the oocyte synthesizes and incorporates many materials into its organisation. Modern biologists believe that the egg establishes the foundations of the future embryo while still in the ovary.

It contains a variety of materials that in some way determines the development of the embryo. It manufactures glycogen, lipids, proteins, and nucleoproteins. It takes up the precursors for future synthetic processes from the environment and stores them in the cytoplasm. The contents are not distributed uniformly within the egg but are arranged in an organised and specific manner in layers or regions. The cytoplasm is thus heterogenous. This stratification of materials within the egg endows it with *a polarity,* since the materials are usually more concentrated at one pole than the other.

The submicroscopic structure of the cortex differs from that in the interior of the cell; the yolk becomes concentrated at one area of the cell more than at another. Different organelles are located in different parts of the cell. The egg with its mosaic of organised substances becomes partitioned through cleavage and gives rise to cells containing different types of cytoplasm as indicated in Figure elsewhere in this chapter, which shows the eight-cell stage of amphioxus. Obviously the cells formed from the yolk-rich *vegetal pole* of the frog's egg contain more yolk than those formed from the *animal pole area.*

The mass of cells resulting from cleavage are not duplicates of one another except genetically, but each cell has its own individuality, since each contains a nucleus, nucleoli, cell organelles, and a specific volume of the original cytoplasm surrounded by a cell membrane. The packaging of the cellular materials does not go on at random, however. Cellular division occurs in such a manner that by the time the blastula stage is reached the cells are grouped together into geographic areas. These areas can be marked, identified, and followed through the subsequent stages of development.

Nontoxic vital dyes or particles, such as carbon, embedded on the surfaces of the cells visualise cellular movement. In some cases advantage can be taken of naturally occurring markers, such as pigment granules. As the embryo develops, the tagged cells migrate to various areas of the body and differentiate into tissues and organs. Eventually, a depletion of reserve cytoplasmic materials brings cleavage to a halt.

The synthetic processes associated with chromosome duplication, spindle-fiber formation, and membrane construction diminish. Each blastomere contains its specific quantity of cytoplasm and the cell attains its adult size.

DISTRIBUTION OF NUCLEAR GENES

Since each new blastomere contains its own diploid number of chromosomes, a doubling of chromosomal material (DNA and accompanying protein) occurs with each cleavage, just as in adult mitosis. The chromosomes contributed by the male as well as those of the female replicate and distribute themselves to each blastomere.

Cleavage is a time of great synthetic activity as far as the nucleus is concerned and results in hundreds of blastomeres, all with identical nuclei. In some animals (sea urchin, for example) synthesis of DNA is one of those processes stimulated by fertilisation. It goes on at the expense of materials stored in the egg. In some invertebrate animals, even before the pronuclei fuse, the pronuclei start their synthetic activities as shown by the uptake of thymidine.

Replication begins during interphase and the chromosomes double before mitosis starts. If the eggs are divided into nucleated and nonnucleated halves by centrifugation, both parts of the egg undergo division. Obviously the nucleus is not necessary for cleavage. In these cases centrioles apparently appear anew and induce furrow formation. Nevertheless, the nonnucleated half does not continue through many divisions, usually not past the blastula stage.

Thus, the nucleus appears to be essential if the zygote is to differentiate into a normal animal. At fertilisation the ratio of nucleus to cytoplasm is abnormally small when compared to adult cells. As cleavage continues, this ratio increases. According to Brachet (1950), in a mature sea urchin egg the ratio $\frac{\text{volume of the nucleus}}{\text{volume of the cytoplasm}}$ is 1/550. By the time the blastula stage is reached, the ratio has changed to 1/6.

The duplication of chromosomes during cleavage gives rise to nuclei that are equivalent in every way. This has been demonstrated many times. In 1928 Spemann devised a simple but elegant experiment to test this idea. He tied hairs around the fertilised eggs of *Triturus* (newts) as they were about to cleave. This famous experiment is illustrated in Figure elsewhere in this chapter. As a result, the cytoplasm of one half of the egg lacked a nucleus but was connected by a narrow bridge of cytoplasm to the half that still contained a nucleus. When the egg nucleus divided, the cytoplasm on that side of the ligature cleaved; the other side did not.

Spemann allowed the cleaved side to reach the 16-blastomere stage and then released the hair loop so that one of the nuclei could pass through the cytoplasmic bridge to the uncleaved side. When this happened, this side also started to cleave. Eventually, Spemann completely separated the two halves of the egg and in several instances obtained two complete embryos. He concluded that the nuclei from a 16-cell stage were similar if not identical to the original zygote nucleus and were capable of directing normal development.

It was inferred that cleaving nuclei of early embryos are completely equivalent to the fertilised nuclei and to one another. The concept of the equivalency (or potency) of nuclei was retested with more modern techniques. Nuclei from cells in different states of development were transplanted into enucleated eggs.

These investigators demonstrated that in an amphibian the nucleus of a differentiated cell retains the ability to direct embryonic development, although it may be highly abnormal. The nucleus of an unfertilised

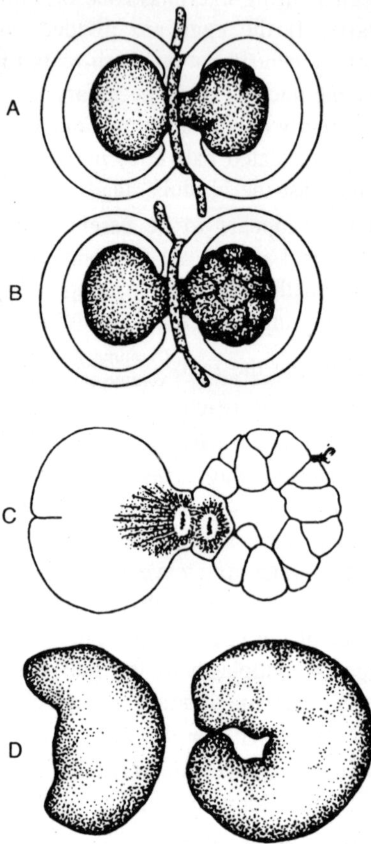

Figure 6.3 : Spemann's experiment proving that the duplication of chromosomes during cleavage gives rise to nuclei that are equivalent. A, Hair tied around a fertilised Triturus *egg separated it into one half with a nucleus and one half without. B, Right half with a nucleus continued to divide, the left side did not. C, At the 16-blastomere stage the hair was released and a nucleus was allowed to pass through the cytoplasmic bridge to uncleaved side. The left side then started to cleave. Twin embryos developed, although development of left embryo lagged behind the right one.*

egg is located near the surface of the egg. One can, with the aid of a microscope, remove it with a needle or (in *Xenopus laevis)* kill it with pinpointed ultraviolet radiation. As a result, one obtains an enucleated egg with intact cytoplasm.

Gurdon then removed the diploid nucleus from a differentiated somatic cell, in this case a cell from the intestine of a feeding tadpole, and inserted this nucleus back into the ennucleated frog's egg. He postulated that if the nucleus of a differentiated cell were combined with egg cytoplasm and brought about normal embryonic development, that is, an embryo containing all cell types, then the nucleus did not lose genes during cell differentiation. In a few cases frogs developed that were normal in all respects.

The development of these normal frogs showed that the specialised somatic cell had, in fact, retained a copy of every different gene and was capable of differentiating cells, not just intestinal epithelium. When inserted into the cytoplasmic environment of an egg, the nucleus of a differentiated cell, like the original nucleus of a germ cell, carried out the various processes involved in development.

Gurdon also made serial nuclear transplantations, using the same procedure described above with one exception. A larva that itself resulted from nuclear transplantation (rather than from fertilisation of an egg) donated the nucleus from one of its differentiated cells. When this procedure was repeated several times, the nucleus of a larva several generations later still retained its ability to direct normal development.

The resulting population is called a *clone, a* population consisting of individuals that are genetically identical. The concept of serial transplantation is illustrated in Figure elsewhere in this chapter.

It is a technique that enables one to test the potency of the genes. If genes are lost or altered in the blastomere nucleus during cleavage, normal development of the egg does not occur. Nuclei obtained from early cleavage stages, the morula, blastula, gastrula, and differentiated organs were tested. Gurdon (1968) removed the nuclei from some specialised cells-intestinal cells with brush borders-from tadpoles that had begun to feed.

He inserted these diploid nuclei into eggs from which the nuclei had been removed. In most cases abnormal development occurred, but a few of the eggs cleaved and grew into normal adult frogs. Spemann showed that nuclei from early cleavage stages (16-cell stage) brought about normal development. Gurdon and Briggs and King carried the problem a little further. They demonstrated that nuclei from differentiated cells in a certain small percentage of cases, retained their ability to

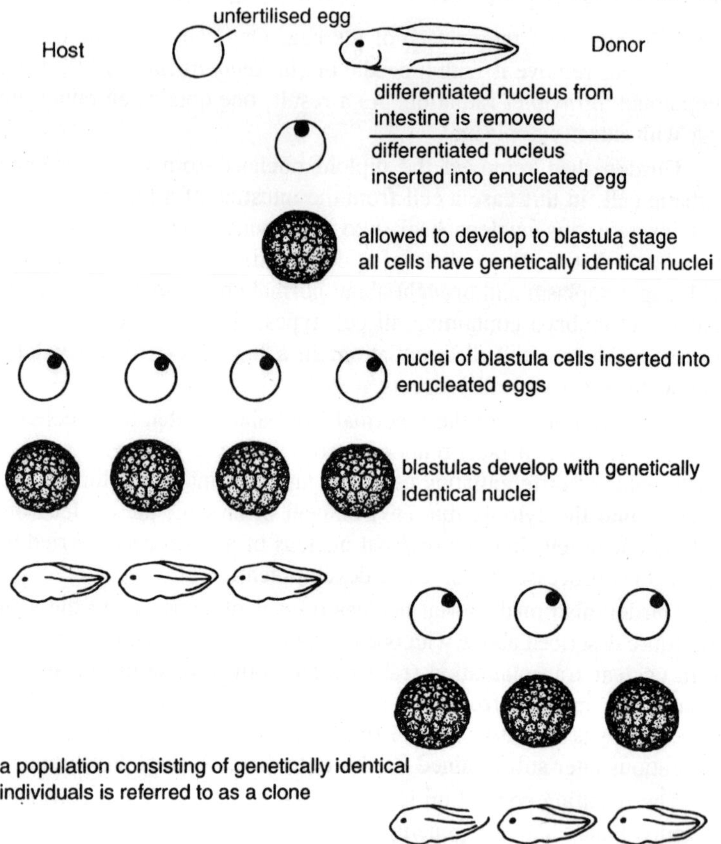

Figure 6.4 : Drawing of serial transplantation.

direct the egg through normal development. In general however, as development continued, the nuclei from the more differentiated cells became progressively restricted. Evidence, then, supports the idea that the genes of the nuclei of early developing cells arc not permanently changed during development, at least when transfers are made *between closely related species* and subspecies. They retain the capacity for directing the development of the zygote to normal maturity.

INTERACTION OF NUCLEUS AND CYTOPLASM

Although the nucleus is an important factor in embryonic growth, the cytoplasm also influences the developmental processes. The interaction between nucleus and cytoplasm plays a significant role. We have seen how the process of cleavage results in the progressive segregation of the cytoplasmic components. The geneticallv identical genes are thus distributed to cells with different types of cytoplasmic components. According to Moore "the action of genes is a function of their location in a specific cytoplasm." The restricting properties of the cytoplasm was particularly evident when he transferred nuclei to enucleated ova of a different species. The arrested development at the blastula stage aptly demonstrated the importance of cytoplasmic influence.

Sally Hennen undertook similar studies but carried the work a little further. She inserted nuclei of *Rana pipiens* into foreign cytoplasm, the enucleated eggs of *Rana sylvatica*. After several apparently normal cleavages, she transferred the *pipiens* nuclei back into enucleated *pipiens* ova. As a result of this back-transfer, she produced abnormal and incomplete development. In most embryos the arrest of development occurred before the midlarval stage.

Careful chromosomal analysis of the back-transferred nuclei revealed abnormal chromosomal numbers and types. The chromosomes had been altered when they replicated in the foreign cytoplasm of *sylvatica*. These chromosomal changes were irreversible and prevented the direction of normal development when returned to their own type of cytoplasm.

FACTORS INFLUENCING CLEAVAGE

Influence of Yolk on Cleavage

Total or holoblastic cleavage

Food is usually stored in the egg in the form of inert yolk granules or platelets. In those eggs in which there is little yolk, dispersion of granules is at random in the cytoplasm. As described earlier, eggs with a homogenous cytoplasm, such as those of amphioxus and mammals, are considered isolecithal.

The sparcity of yolk offers little impediment to the cleavage furrow and the egg divides completely into two blastomeres. Complete or total cleavage is referred to as *holoblastic*. In the tclolecithal eggs of amphibians and certain other vertebrates, the yolk is in the form of large yolk platelets concentrated at one pole, the vegetal pole. Although the dense platelets retard the furrows, they do not inhibit them. This type of egg is also completely divided (holoblastic cleavage).

Two "laws" based on observations of early biologists predict cleavage planes.

1. In most spheroidal eggs the nucleus lies in the center of the protoplasmic mass. The mitotic spindle tends to elongate and orient in the direction of the greatest protoplasmic mass. The plane of cell division divides the long axis of the spindle transversely.

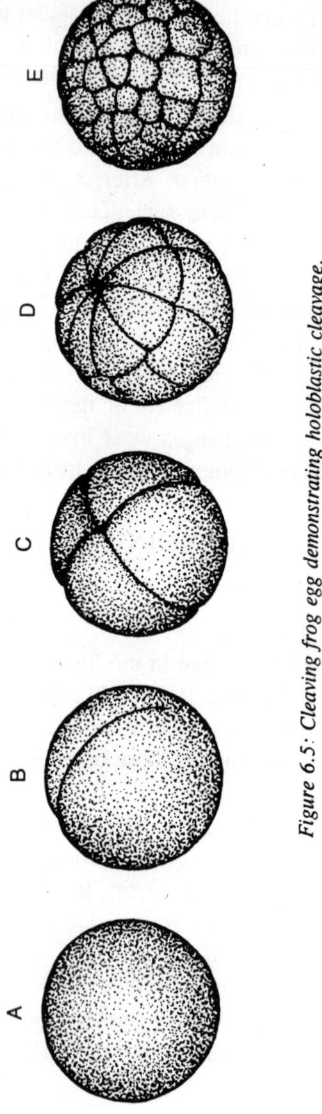

Figure 6.5: Cleaving frog egg demonstrating holoblastic cleavage.

2. The second cleavage "law" states that each cleavage furrow tends to bisect the previous furrow thus dividing the egg into equal parts.

In a typically holoblastic cleavage the first furrow is mcridional, that is, it tends to bisect both poles of the egg and passes through the center of the egg as demonstrated in Figures elsewhere in this chapter. The second cleavage plane is also meridional but at right angles to the first. In general, in this type of cleavage the blastomeres formed are equal, and the second cleavage results in four equal blastomeres.

The third cleavage conforms to the second cleavage law. It is latitudinal; it bisects the median axis somewhere between the two poles. In those eggs with little yolk, the plane of the third cleavage approaches the equatorial plane.

However, even in isolecithal eggs there is usually enough yolk to displace the third cleavage plane slightly toward the animal pole. In the telolecithal eggs of the frog, the unequal distribution of yolk displaces the third horizontal cleavage even further from the equator. As a result of this third cleavage, unequal blastomeres are produced: the larger cells at the vegetal pole are referred to as *macromeres* and the smaller ones at the animal pole, *micromeres*.

The fourth cleavage plane in holoblastic cleavage is again meridional, bisecting the third. Two furrows, passing from pole to pole, synchronously divide the eight cells into sixteen blastomeres. After the fourth cleavage, synchrony is lost in many vertebrate eggs. Blastomeres containing the heavy yolk platelets divide at a slower rate than those at the animal pole. The larger, yolk-laden cells lag behind. While a new cleavage plane bisects the micromeres, the previous furrow completes the division of the macromeres at the vegetal pole.

As a result of continued cleavage, more macromeres than macromeres form. The blastomeres flatten along the surface in contact with one another but round on their free surface. At first they form a closely packed ball held together by the vitelline membrane. Later, the cells adhere to one another on their own.

At this stage the embryo resembles a mulberry and is referred to as the *morula stage*. Narrow spaces appear between the cells after the third cleavage. They eventually coalesce, forming a space in the center of the morula. This cavity increases in size as cleavage continues. The morula is replaced with the next stage of development, the blastula. This hollow ball of cells is composed of a layer of cells, the blastoderm, surrounding the cavity, or *blastocoel*. The blastocoel, resulting from

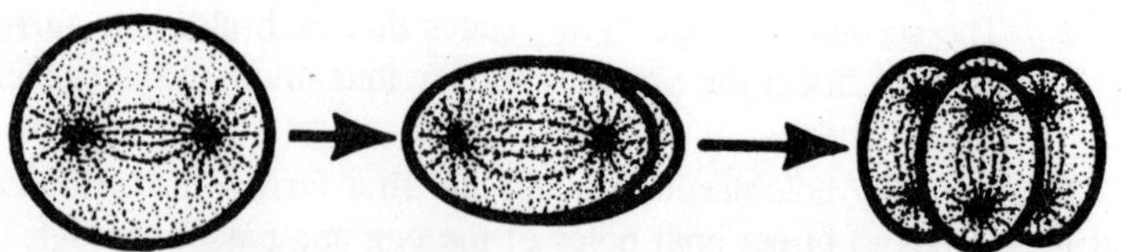
Figure 6.6 : Diagram demonstrating the laws of cleavage. Mitotic spindle orients itself so that its long axis is parallel to that of cytoplasmic mass.

cleavage of isolecithal eggs, is more or less centrally located in the blastula as indicated in Figure elsewhere in this chapter. Cleavage of the telolecithal egg gives rise to an eccentrically located cavity. Heavy yolk-laden cells form the floor of the blastocoel and two to three layers of small micromeres compose the roof. Mammalian eggs cleave in a rather specialised manner.

With the exception of the monotremes, mammalian eggs are isolecithal and cleavage is holoblastic. However, in most animals the first two blastomeres produced are of unequal size. In some mammals these first two blastomeres are destined to form two different types of tissue. For example, in the pig, the smaller blastomere will give rise to the formative cells of the embryo but the larger blastomere will become the trophoblast, or nutritive cells. Unlike the lower animals, cleavage is not always synchronous.

The larger blastomere produced by the first cleavage often divides first, resulting in a threeblastomere stage. Five-, six-, seven-, and even eight-cell stages have been observed. By the 16 cell stage, the morula formed in most mammals contains centrally placed cells (the inner cell mass) and an envelope of smaller outer cells, which will give rise to the trophoblast cells as shown in Figure elsewhere in this chapter. In subsequent cleavages these smaller cells divide at a faster rate than do

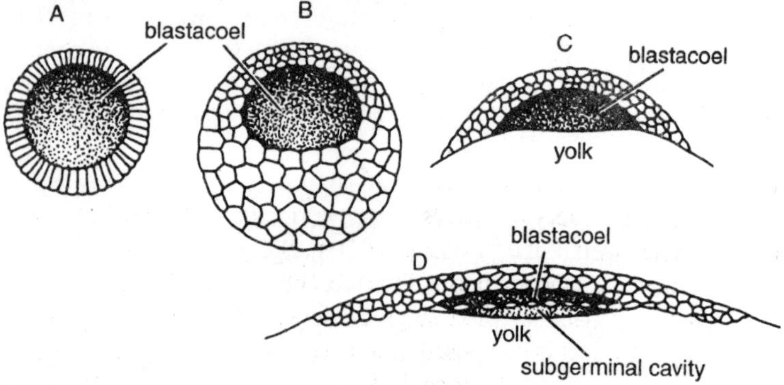

Figure 6.7 : Comparison of, A, amphioxus, B, frog, C, fish, and, D, chick blastulas.

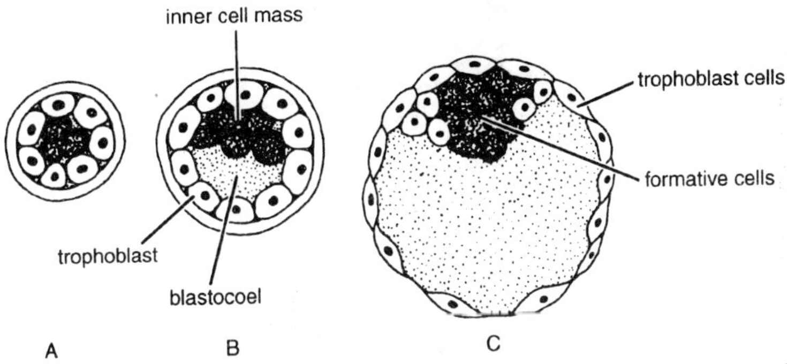

Figure 6.8 : Cleavage of mammalian egg gives rise to formative cells of embryo and trophoblast, or nutritive cells.

the centrally located cells. The blastocoel begins to form by the 32-cell stage.

By the morula stage the embryo enters the uterine cavity and fluid from this cavity passes through the trophoblastic layer, (the outer cells of the morula) and accumulates in the spaces between the cells of the inner cell mass. As a result of the increased fluid, the whole embryo appears bloated and the outer cells become flattened and lifted off the inner cells. This inner cell mass remains attached at only one point to the inner surface of the outer flattened cells. The cavity formed by the surrounding cells is considered the blastocoel. The mammalian embryo is now called the blastocyst.

Incomplete or Meroblastic Cleavage

Further accumulation of yolk occurs in the eggs of sharks, birds, and reptiles. In these eggs the yolk consists of a compact mass surrounded by a thin film of cytoplasm. A disk or cap of cytoplasm *(blastodisk)* forms at the animal pole of the egg and is the only part of the egg that is partitioned. Cleavage is said to be incomplete, or *meroblastic,* when the cleavage furrow cannot bisect the whole egg. The first cleavage furrow is considered meridional in incomplete cleavage.

The furrow bisects the upper portion of the blastodisk but does not divide the more yolk-laden cytoplasm of the lower portion of the disk. In other words, the furrow penetrates only about half the thickness of the disk. The second furrow consists of two furrows, one on each side and at right angles to the first. The third set of furrows is vertical, bisecting the plane of the second cleavage, and therefore parallel to the first; the plane of the fourth cleavage is considered to be vertical also. In Figures elsewhere in this chpater, meroblastic cleavage results

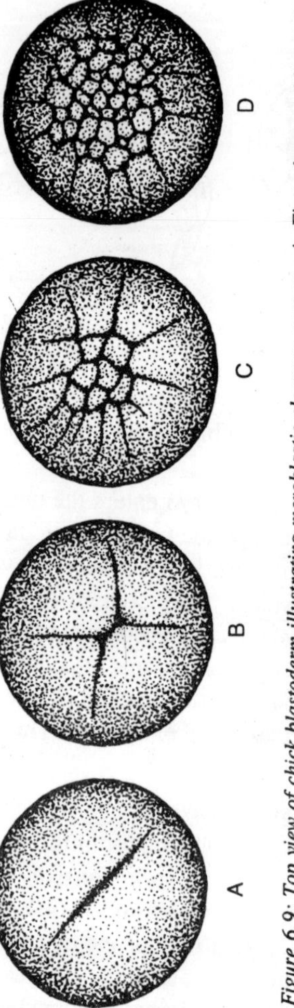

Figure 6.9: Top view of chick blastoderm illustrating meroblastic cleavage pattern. A, First plane meridional. B, Second cleavage furrow forms at right angles to first. C, About 32-cell stage. D, About 64 cells. Note central and marginal cells.

in a cluster of centrally located cells that have top and side surfaces but lack a floor and are continuous with the yolk below. The peripheral or *marginal cells* surround the *central cells*. These cells are incomplete since the furrows do not extend to margins of the disk. They lack a floor and have only two sides (formed by the furrows); their outer edges are continuous with the peripheral cytoplasm.

Up to this point all blastomeres lie in one plane. As cleavage continues, more and more marginal cells develop sides and become part of the central cells. Eventually either a cleft separates the nucleated part of the cell from the rest of the yolk or there is a horizontal cell

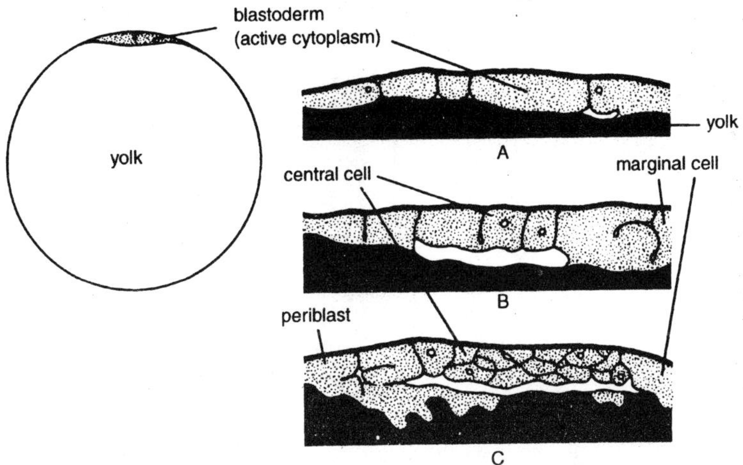

Figure 6.10 : A, Section through blastoderm showing furrow bisecting upper portion of blastodisk and producing central cells with top and side surfaces but open to yolk below. B and C, The marginal cells are continuous with peripheral cytoplasm and yolk; horizontal cleft separates central cells from yolk.

division. In either case, central cells become surrounded by cell membranes and are complete cells, no longer continuous with the yolk. These central cells form a superficial layer above the cleft; on the other hand, the marginal areas and the area beneath the cleft consist of incomplete cells, or *periblast.* In the fishes this superficial layer composes the roof of the cavity, considered to be the blastocoel in these forms.

In most reptiles and birds only some of the central cells are incorporated into the superficial layer, which becomes designated as the *epiblast.* A few blastomeres remain in the space between the epiblast and yolk; these cells later unite into a thin layer, the *hypoblast.* The hypoblast may also form by delamination from the upper layer or possibly by the addition of cells from the posterior thickened area where epiblast joins the peripheral area. In any event, the cavity becomes divided into two parts.

The space between the epiblast and the newly formed hypoblast is referred to as the blastocoel and is considered homologous to that of the amphibian. The area between the hypoblast and the yolk is usually called the subgerminal space. It is the central cells that will give rise to the embryo and the periblast that will function in a nutritive manner. The cavity, which is limited to a space under the central area, causes these central cells to take on a more transparent appearance as compared with the more peripheral cells. These latter cells rest on the yolk and

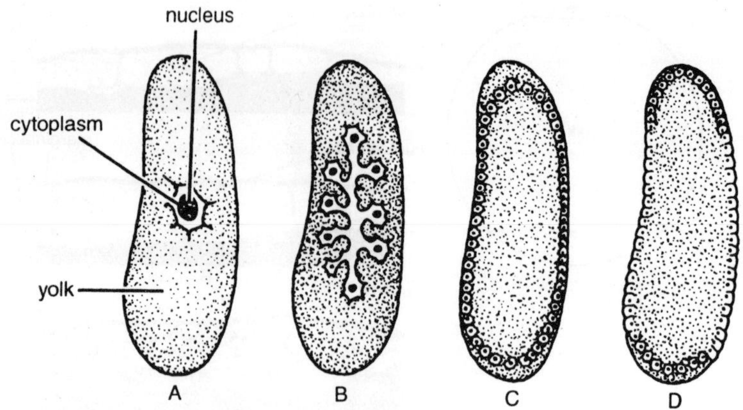

Figure 6.11: Superficial cleavage occurs mainly in insect eggs.

appear more opaque. The blastodisk therefore becomes subdivided into two areas. The centrally located transparent cells constitute the *area pellucida* and the more opaque peripheral area composes the *area opaca.*

Superficial Cleavage

A third type of cleavage should be mentioned although it occurs only in invertebrate eggs, mainly insects and other arthropods. In the insect egg the yolk content is high and is more or less segregated from the cytoplasm. The cytoplasm forms a thin coat around the ovum and surrounds the nucleus in the form of a small island within the yolk-laden insect egg. This type of egg is known as *centrolecithal* and its cleavage, as *superficial cleavage.* The nucleus in the island of cytoplasm first undergoes a number of divisions while the cytoplasm remains uncceaved.

After several mitotic divisions, each nucleus becomes surrounded by a small amount of *cytoplasm and migrates to the slarface of* the egg. A syncytium, consisting of several nuclei in a ring of undivided cytoplasm, develops. After the migration of the nuclei, there are formed many furrows, which pass from the surface inward, dividing the cytoplasm into cells whose inner surface is continuous with the yolk.

Eventually the cytoplasm is separated from the yolk, and the cells, now with complete surfaces, arrange themselves in a layer around the compact central mass of yolk as in Figure elsewhere in this chpater.

Influence of Cytoplasm on Cleavage

Factors other than the amount of yolk deposited in the egg determine pattern of cleavage. The organisation of the cytoplasm itself, which was laid down in the ovary, influences the orientation of the spindle within the egg. This, in turn, determines the type of cleavage that will follow.

Three main categories are usually described: *bilateral, radial,* and *spiral.* Radial and spiral cleavages occur in the eggs of invertebrates. Bilateral cleavage is usually associated with vertebrate eggs.

In bilateral cleavage the spindle is oriented in such a way that the first furrow usually coincides with the median plane of the embryo and separates the right side from the left. In radial cleavage the furrows bisect the egg and are at right angles to one another.

The blastomeres are thus arranged in a radially symmetrical pattern around the polar axis. After the horizontal cleavage, the blastomeres of the upper tier lie over the corresponding blastomeres of the lower tier. In spiral cleavage the spindles become tipped and divisions occur in an oblique direction. As a result the blastomeres of the upper tier shift so that they come to lie over the junction of the blastomeres in the lower tier rather than over the cells themselves.

This is especially obvious at the third cleavage in which an unequal division of cytoplasm occurs, producing micromeres and macromeres. The spiral may turn in either a clockwise or counterclockwise direction, referred to as *dextral or sinistral,* respectively. The coiling of the shells of gastropod mollusks reflect this type of cleavage. The direction of coiling demonstrates the importance of the maternal genes during these early stages of development.

If cleavage spindles are oriented so that division is clockwise (dextral), the shell of the adult snail coils dextrally. On the other hand, the shell coils sinistrally when cleavage occurs in the counterclockwise direction. In one species of snail some individuals are dextral and some are sinistral.

A single gene determines the direction of cleavage in these snails. The allele for dextral coiling (L) proved dominant over the recessive sinistral (1) allele. If the eggs of a sinistral female (11) are fertilised by sperm from a homozygous male (LL), the resulting offspring all cleave sinistrally, despite the fact that the F, genotype is Ll. This cross is diagrammed in Figure elsewhere in this chpater.

If F, individuals are crossed (Ll × L1), the second generation are all dextral. The usual phenotypic segregation does not occur according to normal Mendelian inheritance since three fourths would be expected to spiral according to the dominant pattern and one fourth sinistrally. Actually, the genes do segregate according to normal rules of inheritance (LL, Ll, 11; 1:2:1), but they do not express themselves since the direction of coiling was determined by the *maternal* genotype, in this case Ll.

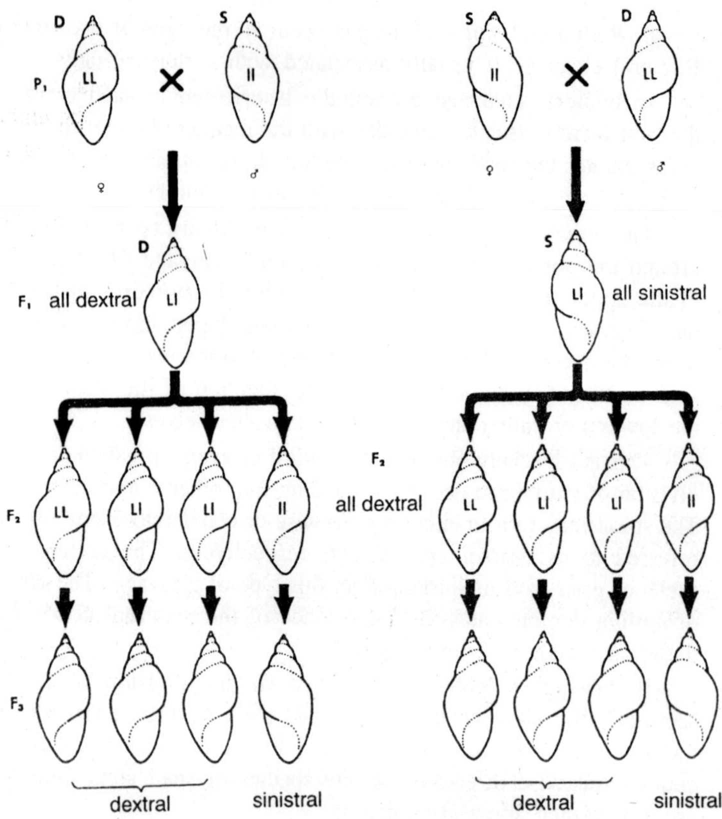

Figure 6.12 : Inheritance of coiling in snails illustrating the influence of maternal genotype.

A different result occurs if the original parents have a different genotype. If the eggs of a dextral female is fertilised by a sinistral male, all of the F_1 snails demonstrate dextral spiraling, dictated by the genotype (LL) of the mother. The coiling in the F_2 generation would also give rise to dextral spiraling as described above.

It is not until the third generation that the threeto-one segregation expresses itself. In this case, eggs obtained from LL or Ll individuals all produce dextral spiraling, but ll genotypes give rise to sinistral. The results of these crosses depend on the organisation of the egg. The establishment of this organisation occurs in the oocyte while the egg is in the ovary undergoing cytoplasmic maturation. The spiraling of the zygote, therefore, is under the control of maternal genes. At the time of fertilisation, the genotype of the sperm does not influence the determin-

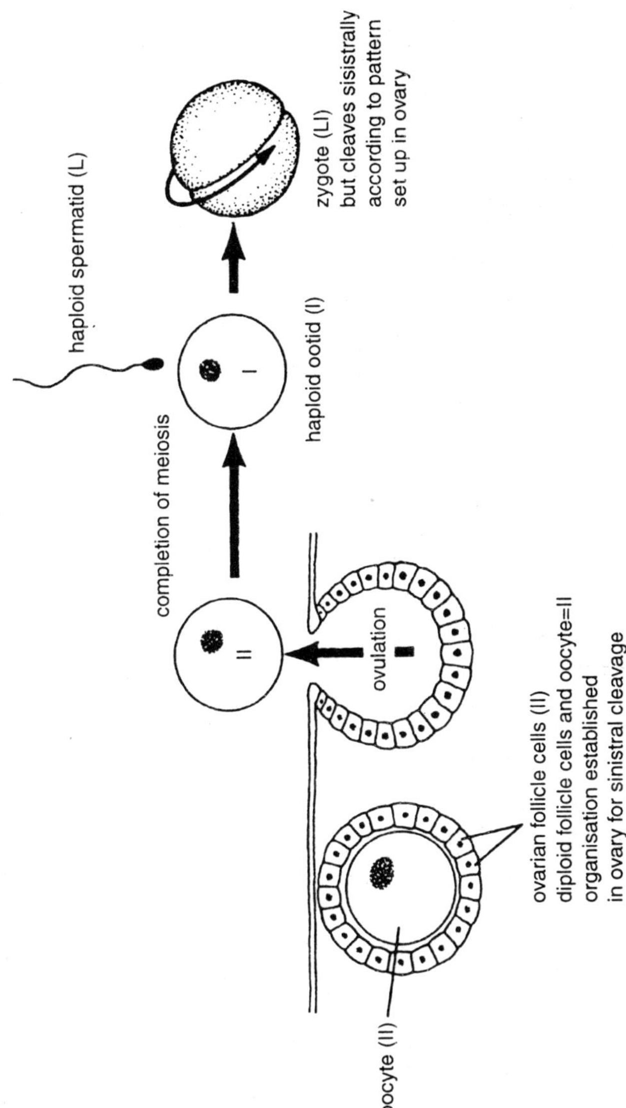

Figure 6.13 : Schematic drawing illustrating the determination of egg organisation in the ovary.

ation of the cleavage pattern. It will be expressed, however, in the next generation.

Rate of Cleavage

In most vertebrate eggs, the first cellular division occurs at a definite rate in relation to the time of sperm penetration. This rate of cleavage is definitely affected by temperature, which influences the rate of all chemical and physical processes. If a batch of frog eggs are fertilised and all are maintained at the same temperature, the time of the first cleavage furrow can be predicted.

For example at 18° C the first cleavage furrow will divide the egg into two blastomeres in about 3½ hours, but at 25° C, the furrow appears in 2½ hours. At 25° C the second furrow forms an hour later, the third forms a half hour later, and the fourth, in 20 minutes. As each cleavage occurs, the time between formation of successive furrows decreases. In general, at each division all vertebrate blastomeres are divided, resulting in a doubling of the total number: 2, 4, 8, 16, 32, etc.

This rule, however, does not hold for mammalian eggs in which, as pointed out earlier, synchrony is lost and cells divide on an individual basis. Although temperature is an important factor, the rate of cleavage is specific for a particular species and is under cytoplasmic rather than nuclear control. Different species have different rates of cleavage and these rates have been programmed into the cytoplasm of the egg synthesised in the ovary.

The fact that the maternal genes determine the rate of cleavage may be demonstrated by fertilizing an egg of one species with the sperm of another species, whose cleavage rate is different. If cleavage occurs in this type of experiment the rate is always characteristic of the mother species.

RESULTS OF CLEAVAGE

Beginning of Vertebrate Body Plan

As stated already that the vertebrate body plan is bilaterally symmetrical, has an anteroposterior axis, and a dorsal ventral surface. This basic plan is established early in development. Just how early seems to depend upon the specific vertebrate under consideration. As indicated already that the synthesis of yolk is a definite feature of cytoplasmic maturation.

The polarity of the egg is usually determined in the ovary and the yolk accumulates at one end of the egg, the vegetal (or vegetative) pole. However, sometimes this localisation of yolk does not occur until fertilisation (such as in many teleost fish). The nucleus of the unfertilised egg (the germinal vesicle) migrates away from the yolk to the opposite end of the egg, marking the animal pole for isolecithal eggs.

The cytoplasm of the egg is not homogeneous. The potential for greater activity rests with the cytoplasm of the animal pole and decreases in gradient fashion to the vegetal pole. Polarity of the egg is usually explained as quantitative or qualitative differences in the cytoplasmic substances of the egg and involves more than just the deposition of yolk. These differences have a definite relationship to the organisation of the embryo and the establishment of the adult vertebrate plan described above.

The area of the animal pole, that region shown to be most physiologically active, usually becomes the anterior end of the embryo, and the material at the vegetative pole lies in the future ventral region of the embryo. Apparently the polarity of the cytoplasm assumes a definite relationship to the future embryonic axis and body organisation. The majority of animals, including all vertebrates, *are bilaterally* symmetrical; that is, they possess only one pair of symmetrical sides.

If a median longitudinal or sagittal plane passes along the anteroposterior axis, it divides the animal into two symmetrical halves that are mirror images of one another. In some cases, the first cleavage furrow divides the egg into right and left sides and establishes the future symmetry of the vertebrate body. Bilateral symmetry is believed to have adaptive value for the animal that moves forward on a substratum. It contributes to cephalisation and helps orient the individual in directional movement.

The question arises, as it does for all vertebrate structures, whether bilateral symmetry preexists in some manner in the egg or whether it depends on the influence of other factors, such as the environment. The subject has been studied mostly in amphibians and birds, and in these animals the plane of symmetry is not definitely preformed in the egg, but is imposed at a later date. The amphibian sperm penetrates the cortex of the egg at any point around its circumference, approximately 20 or 30 degrees down from the midregion of the animal pole.

As it enters the egg, it carries with it a trail of pigment granules from the surface of the egg, marking what is known as the *penetration path.* Since the pigment granules are concentrated in the outer cortical region, *they* can be used to visualize this superficial layer. Extensive movements of the surface layers occurs at the time of fertilisation. The granules are seen to migrate over the animal pole toward the site of sperm entrance as shown in Figure elsewhere in this chapter.

As a result of sperm penetration and cortex migration, in 30 to 60 minutes (depending on the temperature of the water) the area opposite the sperm entrance contains less pigment granules, is gray in colour, and assumes a crescent shape. This area is known as the *gray crescent.*

It is located in the marginal area where the yellowish material of the vegetal pole meets the darkly pigmented cytoplasm of the animal pole. The plane that passes through the point of sperm entrance and through the egg's axis divides the gray crescent into two equal halves and determines the bilateral symmetry of the egg and of the future tadpole. The entrance of the sperm determines the site of the gray crescent, and once the crescent is formed, the plane of symmetry is fixed once and for all.

In the majority of cases it is the first cleavage furrow that passes through the gray crescent and determines the right and left side of the individual. The site. of sperm entrance becomes the future anterior end of the embryo. The material of the gray crescent gives rise to the posterior region. The determination of symmetry does not appear to be rigidly fixed but remains labile during the earliest stages of development.

In the majority of cases the first cleavage furrow divides a urodele egg into right and left halves, but if the first two blastomeres are separated, each blastomere is capable of developing into a smaller but normal embryo with its own bilateral symmery. One can conclude, therefore, that the factors normally responsible for bilateral symmerty are established early in the egg but can be modified by external factors.

The eggs of amphibian, chondrostean fish, and prototherian mammals established bilateral symmetry before cleavage begins by the rearrangement of ooplasmic materials at the time of fertilisation. On the other hand, the eggs of teleosts, birds, and reptiles are retarded in symmetrisation. In these animals with meroblastic cleavage, bilateral symmetry depends on the position of the egg in the uterus, that is, whether the pointed or blunt end faces the cloaca, and how the egg axis orients in space.

According to a recent report the rotation of the egg, as it moves down the oviduct, forces the chick blastodisk into an oblique position; the force of gravity determines the bilateral plane. The anterior end always develops at the lowest point of the blastodisk and the posterior part of the embryo at the highest point.

Regardless of the precise mechanisms involved, authorities usually agree that by the time the area pellucida forms, the final plane of bilateral symmetry is established. The moment when bilateral symmetry is fixed in mammals is uncertain. According to Rugh (1964) after 6½ days of gestation the dorsoventral axis of the mouse develops perpendicularly to the long axis and perpendicularly to the ligament supporting the uterus. The formation of the primitive streak indicates the posterior end of the embryo.

Formation of Blastula

Types of chordate blastulas

As a result of cleavage a cluster of cells forms, the *blastula,* which assumes different shapes depending on the species involved. A diagrammatic comparison of different blastulas is presented in Figure elsewhere in this chpter. If cleavage is holoblastic, the mass of cells takes the form of a hollow ball, but if cleavage is incomplete (meroblastic), cells assume the shape of a disk f blastodisk). Both holoblastic and meroblastic cleavage give rise to blastulas that can be classified into two types.

1. In one type of blastula all the cells become incorporated into the body of the embryo. The zygotes of amphioxus and amphibians form blastulas of this type.
2. Two kinds of cells compose the second type of blastula. One cell type gives rise to the embryo proper and the second type of cell forms auxiliary cells. The latter cells function in implantation, nutrition, respiration, and excretion and play no part in forming the body of the embryo. Elasmo branchs and teleost fishes, reptiles, birds, and mammals possess this type of blastula.

The blastula plays an important role in the development of the individual and represents the first embryonic stage after fertilisation. A blastular stage must be formed before the embryo can follow the sequence of steps leading to the establishment of the normal adult animal. The techniques of vital staining or marking ofblastomeres enables biologists to construct maps of the prospective areas of the embryo.

The labeled cells can be traced as the embryo develops and the fate of various areas followed. The mapping process reveals that certain cells give rise to certain germ layers, which, in turn, make up specific tissues and organs within the embryo. They have a predetermined destiny and under normal conditions give rise to specific major organ-forming areas of the future adult.

These germ layers are the *ectoderm, mesoderm,* and *endoderm,* terms that mean outer, middle, and inner skins and refer to their position within the embryo. Since the cells of the blastula are located on or near the surface of the embryo and as yet do not exist in layers, the area of the future germ layers during the blastula stage is referred to as *presumptive*. For example, that area of the blastula destined to form the neural ectoderm is called the *presumptive neural ectoderm.* The early biologists constructed maps of the blastulas based on the trajectories of tagged cells. They showed the fate of any region in the blastula under normal developmental conditions and were referred to as *fate maps*. Similar patterns exist in all vertebrates. Regardless of the specific shape of the blastula (round or disk-shaped), the organ forming areas are arranged in a pattern around the presumptive notochordal region.

The following general description of four different types of vertebrate blastulas emphasizes the similarity of this basic plan. This pattern is established despite the fact that the eggs contain varying amounts of yolk and employ different types of cleavage patterns.

Protochordate blastula

An understanding of blastulation in amphioxus demonstrates this fundamental developmental process in its simplest form, since the egg of this protochordate contains little yolk. As a result of cleavage the cells arrange themselves in the form of a hollow, pear-shaped structure as shown in Figure elsewhere in this chapter.

The blastoderm forms a single layer around the cavity or blastocoel. Since holoblastic cleavage divides the egg more or less equally (it is never entirely equal), the resulting blastomeres are of approximately the same size. Consequently, the cavity is centrally located within the ball of cells. The cells from the vegetal pole of the egg will become the floor of the blastula, or *hypoblast,* and are the future endoderm of the embryo.

Those cells from the region of the animal pole will form the roof over the hypoblast, the *epiblast,* and will give rise to the epidermis, the neural cells, and notochordal and mesodermal cells. Most of the cells of the composite epiblast will give rise to epidermal ectoderm. Some small epiblast cells with special staining affinities located at the margins of the prospective endoderm layer, form a ring around the blastula.

Figure elsewhere in this chapter shows that this ring of cells consists of two general areas, the dorsal and ventral crescents. The cells of the dorsal crescent give rise to the presumptive notochord and neural plate cells. The ventral crescent contains the cells that make up the presumptive mesoderm.

Amphibian Blastula

The blastula of the amphibian embryo is also a round structure that can be roughly divided into three regions as shown in Figure elsewhere in this chapter. They are similar but not identical to those of amphioxus. The amphibian egg contains a greater amount of yolk than an amphioxus ovum. As a result the cells formed from the vegetative pole of the egg, the hypoblast, lack pigmentation and are much larger than those at the animal pole. They push into the blastocoel causing this cavity to become eccentrically located.

The large yolkfilled cells become the endoderm and will turn up later in the gut region. The rest of the cells compose the epiblast layer, but unlike the blastula of amphioxus, the epiblast of the amphibian is several cells thick. In the frog, the darkly pigmented cells, derived from the animal pole of the egg, give rise to the future ectoderm of the animal.

Further division of the ectodermal regions provide cells for the epidermis and the nervous system of the embryo. The third area of the blastula consists of a ring of cells around the equator. Most of these cells occupying the equatorial region are presumptive mesoderm. Unlike the amphioxus blastula in which the presumptive mesoderm cells occupied approximately half of this marginal zone (ventral crescent), the presumptive mesoderm cells of the amphibian almost encircle the blastula.

The notochord material, wedged between the tapering edges of the mesoderm, can best be seen in dorsal view. It is located just dorsal to those cells that will first migrate into the interior of the blastula. Just above the notochordal cells is a crescent-shaped area, the presumptive neural ectoderm mentioned above. These cells will give rise to the brain and spinal cord of the individual.

Avian blastula

The area pellucida, consisting of the central cells freed from yolk, contributes to the formation of the embryo. It is the part of the chick blastoderm included in the fate map. It consists of two layers of cells/those cells exposed to the surface, the epiblast, and those cells—located below the cleft in the area pellucida, the hypoblast. As in lower animals, the blastula is bilaminar (2 layers).

Instead of having a rounded appearance it may be thought of as being flattened, as if the epiblast of the amphioxus blastula were flattened against the hypoblast, changing the blastocoel into a slitlike cavity as shown in Figure elsewhere in this chapter. The blastodisk (the flattened blastula) rests on a bed of undivided yolk. Since the avian blastula includes the blastomeres located on each side of the cleft, the fate map must treat each layer (epiblast and hypoblast) separately. The formation of the embryo proper involves only a limited portion of the area pellucida.

Some of it contributes to the formation of extraembryonic membranes. Since the shell limits the environment of the developing chick embryo, these necessary membranes function in respiration, excretion, and nutrition-physiological processes that are easily handled when the embryo develops in an aquatic medium. As in the other blastulas studied so far, the marking of the cells during cleavage reveals that the cells of the epiblast are organised around the notochord.

The cells in the anterior part of the area pellucida form the ectoderm and embryonic membranes. The posterior half of the area pellucida gives rise to mesoderm of the body proper and to the extraembryonic mesoderm of the membranes. In the middle of the area pellucida the presumptive neural plate cells take the form of a transverse crescent-shaped mass.

Posterior to the presumptive neural plate cells, near the middle of the area pellucida, lies the presumptive notochord material. The lateral edges of these two areas do not extend to the edges of the area pellucida. In general, the hypoblast, those cells below the cleft, are destined to form the endoderm. The gut incorporates some of these cells into its wall, but most of them will form the lining of the yolk sac. Fig. 6-18, B, is a schematic drawing of an early chick blastoderm sectioned through the median axis in an anteroposterior plane.

Mammalian blastula

It is almost impossible to mark or stain the cells of the mammalian embryo and follow their path through the sequence of developmental phases. The rat, mouse, rabbit, pig, and monkey are the eutherian mammals most studied, and much of our knowledge is based on these forms.

Only a few human embryos arc available for examination, but one can assume that their development parallels that of the monkey. Despite the fact that the mammalian egg undergoes holoblastic cleavage, the cells composing the inner cell mass spread out and form a plate or disk resembling the blastodisk of the bird and reptile. The mammalian blastodisk becomes divided into epiblast and hypoblast as in the chick and reptile, but there is no consensus as to how this occurs.

A flattened layer of cells become distinguished from the rest of the inner cell mass and lines the interior surface of the blastocoel as indicated in the schematic drawing of the blastodisk of the monkey. This layer of cells, considered to be the hypoblast, is formed either by the splitting of cells from the lower aspect of the inner cell mass or by the migration of some of the enveloping cells from the edges of the inner cell mass along the internal surface.

The specific source of these cells varies with the specific mammal

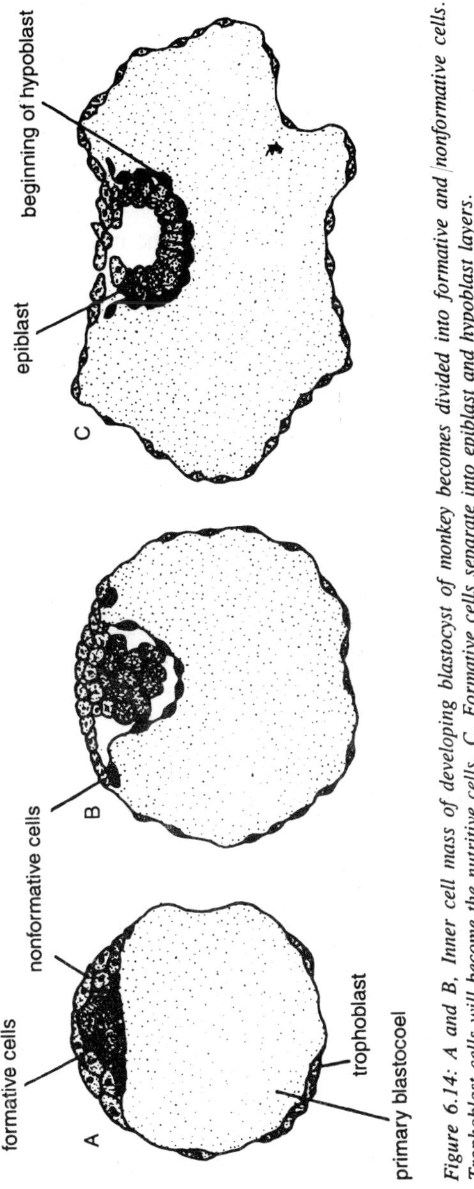

Figure 6.14: A and B, Inner cell mass of developing blastocyst of monkey becomes divided into formative and nonformative cells. Trophoblast cells will become the nutritive cells. C, Formative cells separate into epiblast and hypoblast layers.

involved. Regardless of its origin, the hypoblast will give rise to the endoderm as in other animals.

The cells that remain in the inner cell mass are considered to be the epiblast and will give rise to the epithelium, neural ectoderm, notochord, and mesodermal components of the embryo. The mammalian fate map resembles that of the reptile and bird. The mammalian blastula, however, is composed of cells destined to fill two different functions. Together, the hypoblast and epiblast cells are considered to be the *formative cells* of the embryo, those cells destined to give rise to the embryo proper.

The envelope of cells surrounding the inner cell mass will become the nutritive cells, or *trophoblast*. The lack of yolk in the mammalian egg forces the embryo to find a nutritive source early in development. The trophoblast cells function to establish a relationship with the maternal body-source of the necessary nourishment. The dichotomy of function in the cells of the mammalian blastula can be traced back to the first cleavage division, although little qualitative difference can be determined between these first two blastomeres.

In fact one of the first two mouse blastomeres can be destroyed, and under certain controlled conditions the other will develop into a smaller but normal mouse. The fact that the two blastomeres can be separated and each give rise to a complete individual indicates that all the nuclear and cytoplasmic factors necessary for normal development are contained within each cell. As discussed already that a blastomere that is capable of developing into a complete individual is said to be totipotent.

Significance of blastula stage

There is great uniformity in the early stages of development of vertebrate animals. The essential similarities of gametogenesis and fertilisation have been stressed. The blastula and gastrula are the next stages in development that are fundamentally similar in all animals. Actually one stage merges into another in a continuous process, but it is convenient to classify them for clarity of discussion.

A summary of the steps leading to the formation of the blastula can be enumerated; the particular shape of the blastula is not important:

(*a*) The cytoplasm is cleaved by a series of rapid mitotic divisions into a number of small blastomeres.

(*b*) The packaging of the heterogeneous cytoplasm results in no two blastomeres necessarily having an identical cytoplasmic ceontent.

(*c*) At each cleavage the chromosomes with their genes are replicated and distributed to the balstomere. The nuclei of all the balstomeres are identical.

(*d*) General areas of the blastula are destined to give rise to certain structures, and the segregation of these *major organ-forming areas* is an important function of blastulation.

(*e*) Regardless of whether the areas are to form internal or external organs, most of the blastomeres are arranged on the surface of the embryo surrounding a cavity, the blastocoel.

(*f*) For comparative purposes the blastula is considered to be a two-layered structure composed of an upper epiblast and a lower hypoblast.

(*g*) The major organ-forming areas become arranged in a particular pattern that apparently facilitates their rearrangement during the next step in development.

(*h*) By the time the late blastular stage is reached a physiological center, which will organise and direct the migration of the major organ-forming areas, is established.

(*i*) The rearrangement of the blastomeres through morphogenetic movements (gastrulation) will give rise to the three germ layers out of which the future organs are constructed.

7

Cells in Cancer

The study of cancer cells has been a major part of cell biology for the past 40 years. From this study much has been learned not only about the properties of cancer cells but also about the properties of normal cells. This chapter presents an overview of this research, focusing on the origin and properties of cancer cells. In addition some information is included about practical aspects of cancer as a disease, including what is known about its causes, how it is treated, how it kills, and how it may be prevented.

Cancer is an ancient disease. It has afflicted our ancestors throughout human history and throughout human evolution. Egyptian medical tracts 3500 years old describe diseases today recognisable as cancer. More direct evidence of cancer's antiquity is seen in distinctly scarred mummies and skeletons of ancient cancer victims. Cancer has always been a much feared disease but, until recently, was seldom encountered.

In the last century or so, the situation has changed dramatically. Cancer now ranks second only to heart disease as the major cause of death in the United States and Europe and is similarly devastating in the rest of the world. During 1987, in the United States alone more than 965,000 people developed cancer and about 483,000 died of the disease.

At least one out of every four Americans alive today will develop at least one cancer; nearly 25 percent of all Americans will be killed by cancer. Reasons for the substantial rise in the incidence of cancer are now understood. First, many years of exposure to cancer-causing agents (*carcinogens*) are usually required to produce cancer. The average life span of Americans and Europeans has increased dramatically during the last 100 years, and therefore many more people now live long enough to develop cancer. Half of the people who die of cancer are over the age of 65. Second, modern life-styles and the contemporary environment result in longer exposures to more kinds of carcinogens and in greater amounts. Carcinogens, how they convert normal cells to cancer cells, and how cancer cells differ from normal cells are discussed in this chapter.

CANCER CELLS AND DISEASE

Cancer, like all organic diseases, results from the misfunctioning of one of the many cells that make up the body. Cancers generally begin when a single normal cell is converted into a cancer cell, a conversion that is now partially understood. Each descendant of the single cancer cell is also a cancer cell and all, in turn, produce more cancer cells. Whatever the nature of the conversion, it is transmitted from one cell generation to the next at each cell division.

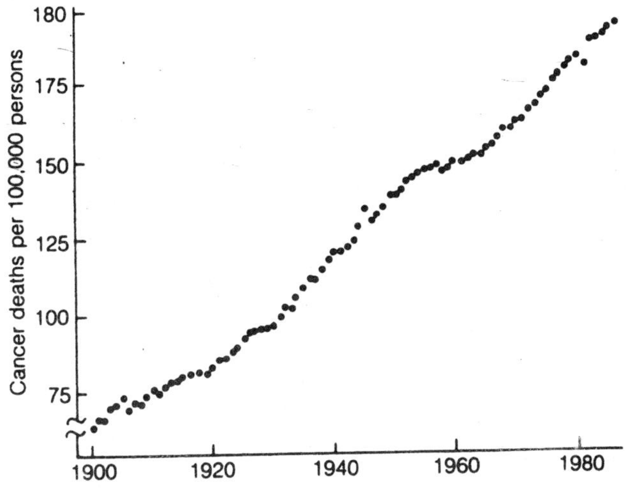

Figure 7.1: Total deaths from cancer per 100,000 persons in the United States from 1900 to 1986.

The misfunctions characteristic of cancer cells are fundamentally different from the misfunctions that occur in other diseases. In almost all other diseases, the misfunction stems from the injury or death of normal cells. Bacteria that cause tuberculosis, for example, destroy lung cells, and bacteria that cause diphtheria produce a poison that kills cells of many tissues by blocking protein synthesis.

Similarly, polio viruses destroy nerve cells, and hepatitis viruses destroy liver cells. Cancer cells, by contrast, are not injured, nor are they dying. They are, by most criteria, remarkably healthy. Cancer cells have three important properties that underlie the nature of the diseases they cause:

1. *Regulation of reproduction is defective in cancer cells.* In an adult organism the reproduction of normal cells is tightly geared to replace cells that are lost. An insidious property of cancer cells is that they reproduce incessantly rather than responding to signals flat turn off the reproduction of their normal ounterparts.
2. Cancer cells in an organism are intrinsically immortal. Many kinds of normal cells in an intact animal have limited life spans (lymphocytes, intestinal epithelial cells, and epidermal cells of skin, are clear examples). Normal cells with limited life spans give rise to cancer cells with unlimited life spans. Cancer cells do sometimes die, for example because of lack of nutrients in cancerous masses inadequately provided with blood vessels. (As a cancer increases in size, blood vessels grow into it, but the process is sometimes inefficient so that cells deeper in the cancer die).
3. *Differentiation of cancer cells is defective.* Many kinds of normal cells in an organism undergo turnover. Cells die and are replaced by cell division. In some tissues this occurs by reproduction of undifferentiated stem cells. For example, the cells in the outer layer of epidermis die and are shed from the surface of skin. New cells are continuously supplied by reproduction of cells in a layer of undifferentiated cells located at the inner base of the epidermis. This is called the *basal layer* or *germinal cells*. Undifferentiated, reproducing cells that supply new cells to a tissue are called *stem cells*.

On the average when a stem cell in the basal layer of epidermis divides, one of the daughter cells is forced out of the basal layer. Continuous reproduction of the stem cells causes cells in the overlying region to be gradually driven toward the surface of the skin. When a cell leaves the basal layer it becomes permanently blocked in GI and starts to differentiate.

Differentiation is predominantly the synthesis of the protein keratin, which is organised into intermediate filaments. Cells packed with keratin filaments give the outer layer of skin its tough, protective character. The keratin-packed cells die and eventually are sloughed from the surface of the skin. The entire course of events from exit of a cell from the basal layer, through differentiation, death, and sloughing from the surface occurs in a matter of days. Cancer arises in skin when a stem cell in the basal layer undergoes a particular genetic change, in most cases caused by ultraviolet light of sunlight. Cells that leave the basal layer continue to reproduce instead of becoming permanently arrested in G1. They may undergo some degree of differentiation but become blocked in an immature state.

Undifferentiated stem cells are clearly evident in some other tissues, such as intestinal epithelium and bone marrow. In some other renewing tissues stem cells are not readily identifiable. For example, renewal of hepatocytes in liver appears to occur by division of fully differentiated hepatocytes instead of undifferentiated stem cells. However, in liver cancer the cancer cells may resemble hepatocytes, but they are incompletely differentiated and do not function normally.

A cancer begins with a single cell that has become defective in the three characteristics just described as a result of one or another change in its genetic makeup. The genetic changes that produce cancer cells are only partially understood; they appear to consist of mutations or changes in expression of genes involved in the control of the reproduction, differentiation, and life span of a cell.

A typical human cancer cell growing in culture divides about once every 24 hours. At that rate, a single cancer cell would generate over 10^9 descendants in 30 days, a mass that would weigh less than one gram. In another two weeks, the population would increase to nearly 10^{13} cells and weigh several kilograms. Fortunately, cancer cells rarely reproduce nearly so rapidly in a body.

Normally, a cancer requires months or even years to give rise to a cancer mass that produces disease symptoms. Cancer represents an accumulation of cells, but cancer cells do not usually reproduce at particularly high rates. Some types of normal cells, for example blood-forming cells of the bone marrow and epithelial cells that line the intestines, reproduce much more rapidly than almost all kinds of cancer cells. However, because of their immortality and failure to stop reproducing, cancer cells accumulate. Accumulation of these functionally defective cells results in disease.

Spread of Cancer Cells

The immortality of cancer cells, added to their incessant reproduction, results in inexorable accumulation of incompletely differentiated, functionally useless cells that kill the organism by interfering with the functioning of normal cells. The malignancy is enhanced by invasion of the cancer cells into adjacent normal tissue and by the spread of cancer cells to distant sites, where they establish new centers of cancer growth. Spreading occurs by detachment of cancer cells from the primary growth and travel in the blood and the lymphatic system, a process called *metastasis*. In laboratory mice, breast cancers that had grown to about 3 grams were found to release several million cancer cells a day into the blood.

However, few of these cells survived to establish secondary cancer growths, or *metastases*, elsewhere in the animal. Even when particularly malignant cancer cells were injected into the blood of mice, only one in 1000 of those was still alive two weeks later. The immune system probably destroys many wandering cancer cells, but other, unknown protective mechanisms may also be at work. Metastases are ultimately established, often in liver, bone, lungs, brain, and lymph nodes. Metastases greatly complicate the problem of treatment and are often responsible for death.

Immediate Causes of Death from Cancer

How cancer cells kill an individual cannot always be precisely determined, but the general mechanisms are known. About 25 percent of all cancer deaths are the result of interference with the function of an essential organ by the growing mass of cancer cells; commonly the function of lungs, liver, brain, or kidneys is impaired. Ten percent of cancer deaths result from starvation.

About seven percent of deaths are the result of hemorrhage, a frequent cause of death in leukemia. Almost 50 percent of all cancer deaths come from infection. Bacteria, molds, and other infectious organisms that are ordinarily destroyed by the immune system of a healthy individual can form fatal infections in cancer victims. The lowered resistance to infection results partly from the generally weak condition of an individual with advanced cancer but is usually more the result of an impaired immune system. By mechanisms that are poorly understood, many cancers severely suppress the immune system.

Treatment of Cancer

Without treatment, virtually all cancers are sooner or later fatal. The objective in the treatment of cancer is elimination of all cancer cells from an individual. Elimination of every cancer cell is important because cancer can regrow from a single surviving cancer cell. The three traditional methods of treatment are to remove cancer cells by surgery, to kill cancer cells with radiation, or to kill or prevent proliferation of cancer cells with chemotherapeutic drugs or hormones. These treatments are often used in combination. Surgical removal of cancerous tissue is frequently followed by radiation therapy or chemotherapy with the objective of killing any cancer cells not eliminated by the surgery.

All current strategies for treating cancer are limited by the fact that, by the time the disease is discovered clinically, most cancers already consist of billions of cells and usually have already metastasised to other sites. To remove, inhibit, or destroy every last cell of a cancer is difficult and usually impossible with presently available methods.

Development of new ways to treat cancer is centered on manipulation of components of the immune system to kill cancer cells that may be disseminated in various parts of the body. The immunological approach has not yet yielded a primary therapy. A potentially powerful form of immunotherapy is to use monoclonal antibodies to deliver cytotoxic agents specifically to cancer cells.

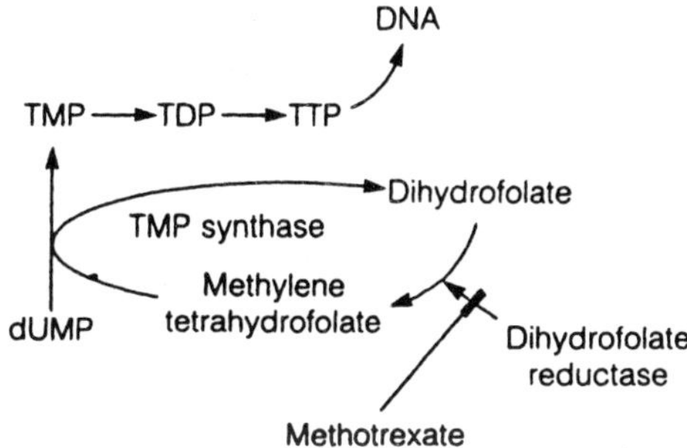

Figure 7.2: Methotrexate is a cancer chemotherapeutic drug that blocks DNA replication by inhibiting the enzyme dihydrofolate reductase.

This immunological approach is based on the putative presence of antigens on the surfaces of cancer cells that are not present on normal cells. For example, in a pioneering test, 104 patients with liver cancer were injected with antibodies tagged with radioactive iodine (^{131}I). The antibodies were specific for an antigen on the cancer cell in the liver.

Fifty of the 104 patients showed decreases in tumor size, and seven appeared to be cured. Liver cancer is ordinarily fatal. In general, the difficulty with the antibody approach has been in finding antigens that are present on cancer cells but not on any type of normal cell. Without such specificity antibodies will also bind to one or another kind of normal cell, producing unacceptable side effects of treatment.

About 35 drugs are now in use to kill cancer cells. Among those frequently used are methotrexate, fluorouracil, and hydroxyurea. Fluorouracil is converted to fluorodeoxyuridine (FUdR) in the cell.

These three drugs block DNA replication as shown in Figures elsewhere in this chapter. Cells in the S phase are killed by such drugs. Other drugs inhibit other cellular reactions or damage cells in other ways, particularly cells that are proliferating.

Unfortunately, none of the drugs is specific for cancer cells and therefore kill normal cells, particularly proliferating cells in the bone marrow that differentiate into blood cells and stem cells in the intestinal epithelium. Nevertheless, chemotherapy is useful in treatment of a limited number of kinds of cancers. One of the most effective uses of chemotherapy consists of treatment of acute childhood leukemia with methotrexate and x-irradiation.

Well over half of treated children with this disease are now cured. Unfortunately, all chemotherapeutic agents are toxic and many are mutagenic. Occasionally cancer patients are killed by chemotherapy. Generally people treated for cancer with chemotherapeutic drugs have an elevated risk of ultimately developing a second cancer induced by the mutagenic action of the drugs used to kill the first cancer.

The overall effectiveness of current methods of treatment can be judged from the mortality figures.

Twenty-three percent of all deaths in the United States in 1987 were due to cancer. Currently, nearly 75 percent of those who develop cancer ultimately die from the disease and the overall cure rate has not changed in the last 30 years. The improvements in cure rate for acute childhood leukemia and a few other cancers do not change the overall cure rate because these cancers account for less than one percent of all cancers. A major improvement in treatment is reflected in a prolonged survival time (but not cure) for many cancer patients.

ORIGIN OF CANCER

The conversion of a normal cell to a cancerous one occurs in at least two steps, called *initiation* and *promotion*. Initiation consists of mutation in or rearrangement of one, or possibly more than one, gene belonging to a class of genes known as *protooncogenes*, which are discussed in a later section. Promotion is less well understood but consists of a second change, not necessarily mutational, that converts an initiated cell into a cancer cell.

This two-step process is in part responsible for the long delay, usually many years, between the start of exposure to carcinogens and development of a cancer. A third, even less well defined process called *progression* may occur in which cancer cells undergo further changes that increase their invasiveness and rates of reproduction and metastasis.

Some cancer-causing agents, X rays for example, can act both as initiators and promoters and are called *complete carcinogens*. Other agents, such as saccharin, appear to work only as promoters.

Causes of Cancer Cell Formation

The causes of cancer cell formation fall into three classes: radiation, chemicals, and viruses. In humans, chemicals are responsible for most cancer. Only several percent of all cancers are caused by radiation and viruses.

Radiation

It has been known for 80 years that radiation can cause cancer, but only gradually have we learned to appreciate how sensitive cells are to complete transformation by radiation. The increasing use of radiation for medical and industrial purposes has been paralleled by increasing rates of cancer in certain; fessions. Radiologists, for example, used to suffer a seven- to ten-fold higher rate of cancer than physicians in general because of the chronic exposure to low levels of X rays. Protective measures have since led to prevention of these excess cancers in radiologists.

Uranium miners have a higher rate of lung cancer than of kinds of miners not in contact with radioactivity. The survivors of the atomic bombs dropped on Hiroshi and Nagasaki in World War II have higher rates cancer. It has been estimated that as many as 39,000 people in Europe, outside the Soviet Union, may eventually die of cancer induced by radioactive fallout fry the failure of the nuclear reactor at Chernobyl, USSR.

There are many examples of cancers induced by X-rays and radioactivity. Tissues particularly susceptible to these radiations are breast, bone marrow, and thyroid gland, although radiation can cause cancer invirtually any tissue. X rays and radioactivity cause cell transformation by mutation or chromosomal re-arrangements of genes involved in regulation of cell proliferation and differentiation. Ultraviolet light is mutagenic, which probably counts for a major part of its action in causing skin cancer.

However, overexposure to ultra violet light also causes suppression of the immune system, and the immune system has a major role in preventing the development of cancer. For example, success rate in transplantation of cancer cells from one animal to another in the laboratory is greatly increased by suppressing the immune system in recipient animals by irradiating them with ultraviolet light before transplantation.

Altogether in the United States radiations from ious sources, but mostly ultraviolet light, probably count for about three percent of all cancer deaths.

Chemicals

Several methods are used in trying to identify chemical carcinogens. First, observation of man populations sometimes reveals carcinogens. Such was the case in the identification of tobacco smoke as the cause of most lung cancer and benzene as a cause of some cases of leukemia in adults.

Second, a chemical suspected of being a carcinogen is administered to animals, and the subsequent cancer incidence is compared with the incidence in animals not exposed to the test chemical. Such tests take years and are expensive. Third, since cancer is the result of a genetic change in a cell and since mo carcinogens have proven to be mutagens, chemicals a tested for their mutagenic properties.

Any chemical thekt proves to be a mutagen becomes a suspected carcinogen. Various methods for testing whether a substance is mutagenic have been developed. The oldest and most widely used, called the Ames test after its originator, is based on mutation in the bacterium *Salmonella typhimurium.* Strains of *S. typhimurium* have been developed that carry mutations that make the cell incapable of synthesizing histidine. Such histidine-minus mutants proliferate in medium containing histidine but not in medium that lacks histidine. Histidine-minus mutants may regain the ability to synthesize histidine as a result of a second mutation, called a back mutation.

Bacteria that sustain such back mutations can proliferate in medium lacking histidine. A substance being tested for mutagenicity is added to a culture of histidine-minus bacteria in a medium lacking histidine. The number of bacteria that regain the ability to proliferate in the histidine-minus medium gives a measure of the mutagenic potency of the test substance. An example of the outcome of such a test is shown in Figure elsewhere in this chpater.

The Ames test is extraordinarily sensitive, takes only a few days, and is inexpensive. Any substance that is mutagenic in the Ames test becomes a suspected carcinogen; however, other evidence, for example animal testing, is required to prove carcinogenicity. Short-term tests like the Ames test provide a quick, inexpensive means of identifying chemicals that should be subject to the far more expensive and laborious tests for carcinogenicity with animals. Most human cancer is caused by chemicals. These chemical carcinogens occur primarily in air, water, and food. The major source of airborne carcinogens is tobacco smoke, which accounts for about one-third of all cancer deaths. Tobacco smoke contains over 6000 different kinds of organic molecules. About 100 of these are known mutagens and about. 30 of these have been tested and shown to cause cancer in animals.

Tobacco smoke also contains minute amounts of radioactive heavy metals absorbed from the soil by tobacco plants. These emit a radiation and include, particularly thorium (^{228}Th), radium (^{226}Ra), and lead (^{210}Pb), which are thought to contribute to the carcinogenicity of tobacco. Carcinogens in tobacco smoke are not only absorbed by cells of the respiratory tract, causing oral, throat, and lung cancer, but also enter the bloodstream and are distributed throughout the body.

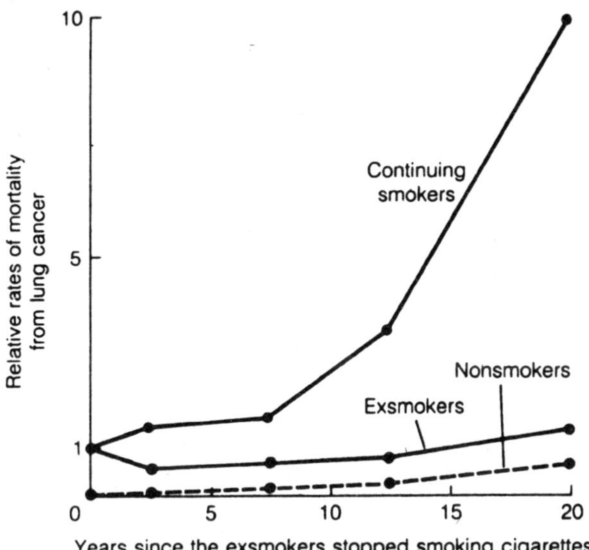

Figure 7.3: Relative rates of lung cancer mortality among nonsmokers, exsmokers, and continuing smokers of cigarettes for a 20-mar period.

Cigarette smokers also suffer from higher risks of esophageal, bladder, kidney, and pancreatic cancers. Pipe and cigar smokers have the same elevated rate of oral cancer as cigarette smokers and have an elevated risk of lung cancer, but not as elevated as in cigarette smokers. Users of smokeless tobacco like chewing tobacco or snuff have an elevated risk of oral cancer.

Altogether, tobacco currently causes about 160,000 cancer deaths per year. Lung cancer is the most common cancer caused by tobacco smoke. Treatments are not effective. About ten percent of those diagnosed with lung cancer survive five years. The discontinuance of use of tobacco has been clearly shown to be followed by a decrease in occurrence in lung cancer. Industry and motor vehicles are minor sources of airborne carcinogens compared to tobacco smoke.

Regular intake of alcoholic beverages carries a higher risk of esophageal and perhaps bladder cancer. Even moderate use of alcohol increases the risk of breast cancer. Tea, including herbal teas, and coffee contain mutagenic substances, but whether they cause cancer in humans is still an unsettled question.

According to various estimates, between 30 and 50 percent of cancers are caused by agents in food. Some foods contain mutagens and carcinogens naturally. Examples are bracken fern (several chemicals), mushrooms (hydrazines), black pepper (piperine), alfalfa sprouts (canavanine), and cottonseed oil (gossypol).

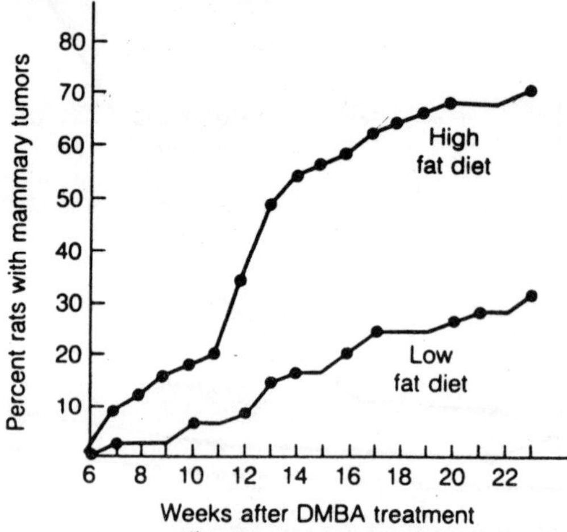

Figure 7.4 : Rats given the carcinogen dimethylbenzanthracene (DMBA) develop mammary tumors more often when given a high-fat diet than when given a low-fat diet.

Some foods are contaminated with carcinogens known as mycotoxins produced by mold that can grow on the food during harvesting and storage. A serious example is the mycotoxin called *aflatoxin* produced by the mold *Aspergillus flavus,* which commonly grows on raw peanuts, corn, grains, and nuts.

Aflatoxin is soluble in lipid and is readily absorbed in the lipid component of these foods. In laboratory animals a diet containing just several parts per million of aflatoxin results in an elevated rate of cancer. Aflatoxins are at least partially responsible for the high rate of liver cancer in Africa.

Other carcinogens are added to foods during processing, especially as a result of pickling, smoking, and fermenting. Chemical reactions that yield mutagens occur in many foods, particularly meats, during cooking at temperatures of 150°C and above.

The most important source of carcinogens in food is probably fat. Animals fed diets rich in fat have higher rates of cancer than animals fed diets low in fat. In human populations throughout the world mortality from cancers of the colon, breast, prostate, and pancreas are correlated with the amount of fat in the diet. Several theories have been proposed to explain how fat might contribute to cancer. For example, the oxidation of fat molecules in a cell results in formation of substances called *free*

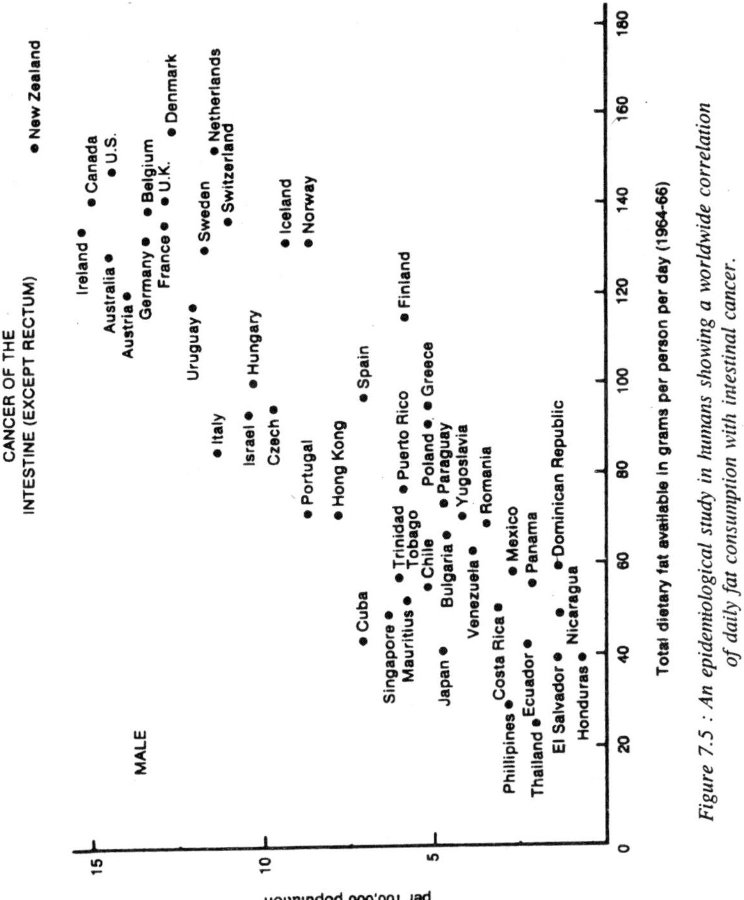

Figure 7.5 : An epidemiological study in humans showing a worldwide correlation of daily fat consumption with intestinal cancer.

radicals. Free radicals possess an unpaired electron, which makes them extremely reactive and mutagenic. Free radicals have extremely short lives, usually measured in fractions of a millisecond.

Several occupations entail exposure to particular carcinogens. Workers engaged in the manufacture of some kinds of dyes have high rates of bladder cancer. Wood workers and leather workers have higher rates of nasal cancer as a result of breathing wood and leather dust. Asbestos workers have a higher risk of lung cancer.

Oncogenic viruses

Oncogenic viruses (onto = tumor; *genic* = forming) have been shown to cause at least a few kinds of human cancers: adult T-cell leukemia, some liver cancers, and Burkitt's lymphoma. Viruses have been implicated but not proven as causative agents in cancers of the uterine cervix, the nasopharynx, and a few other tissues.

However, overall, viruses appear to account for a very small fraction of human cancers. Hundreds of different oncogenic viruses have been identified in animals, primarily in birds and mammals. Usually a virus that causes cancer in one species of animal is harmless in another, although there are exceptions.

For example, Rous sarcoma virus can cause cancer in at least several kinds of birds and in mice. An enormous amount of research has been and is being done on how oncogenic viruses cause cancer, using particularly the transformation of normal cells in culture to cancer cells by oncogenic viruses. This research has led to major new insights not only about transformation by viruses, but also by chemicals and radiation. Only the barest outlines of this work are included here and make up the latter part of this chapter.

REPRODUCTIVE PROPERTIES OF TRANSFORMED ANIMAL CELLS IN CULTURE

Certain properties of normal animal cells in culture are different from those in transformed cells. The conversion of a normal cell to a cancer cell can be achieved in culture with chemical carcinogens, radiation, or oncogenic viruses in a process called *transformation*.

Absence of Density-Dependent Inhibition of Reproduction

Normal cells in culture grow as a monolayer, and only cells at the periphery of the monolayer reproduce. This can be shown by adding ^{3}H-thymidine to the culture medium and detecting DNA replication by autoradiography. Cells at the periphery of a colony become labeled;

Table 7.1: Comparison of Growth Characteristics of Normal and Transformed Cells in Culture.

Property	*Normal Cells*	*Transformed Cells*
Density-dependent inhibition	Present	Absent
Anchorage dependence	Present	Absent
Requirement for serum growth factors	High	Low or absent
Lifee span	Finite	Infinite
G1 arrest and entry into GO	Present	Absent
Karyotype	Diploid	Aneuploid

cells in the center of the colony are arrested in G 1 by density-dependent inhibition. The behaviour of transformed cells is very different. Feeding of ^{3}H-thymidine for 24 hours results in labeling of all cells throughout a colony. In contrast to normal cells, transformed cells in the central part of a colony are not subject to density-dependent inhibition of the cell cycle. When a transformed cell arises in a monolayer of normal cells in a culture, it continues to proliferate, and progeny pile up on one another into irregular layers.

Absence of Anchorage Dependence

Normal cells must attach to a surface in order to proliferate. Indeed, when kept in suspension in nutrient medium, they die within a few days. Transformed cells can usually be adapted to proliferate in suspension culture, that is, they are *anchorage independent*. This property is commonly induced so that transformed cells can be identified by the following method. Cells are seeded in nutrient medium that has been made into a loose gel by the addition of agar. The cells remain stationary in the agar gel and cannot attach to a surface. Normal cells fail to proliferate, and transformed cells form clones visible without a microscope, which can be removed and cultured by standard methods.

Immortality of Transformed Cells

Normal mammalian cells have finite life spans in culture; typically they divide between 50 and 100 times and die. In culture, transformed cells proliferate indefinitely, like their cancer counterparts in an organism. Some have divided thousands of times with no reduction in vitality or reproductive rate.

Absence of the G0 State

Most kinds of transformed cells are unable to enter the G0 state. When subjected to a treatment that arrests normal cells in G1 and causes entry into G0, for example deprivation of an essential nutrient, transformed cells cease proliferation and soon die.

Aneuploidy

Normal cells in culture have diploid sets of chromosomes, that is, they are *euploid*. Transformed cells have an abnormal number of chromosomes, that is, they are *aneuploid*. Typically, the chromosome number in an aneuploid cell is greater than the diploid number. For example, the chromosome number in the human carcinoma cell known as HeLa typically ranges from 55 to over 100.

The HeLa cell was derived from a uterine (cervical) cancer in the early 1950s, and cultures of this cell have been used in research since that time in hundreds of laboratories throughout the world. It is named for Henrietta Lacks, who died from the cancer.

Cancer cells in an animal are also usually aneuploid. In addition, in aneuploid cells extensive exchanges of segments between chromosomes occur so that normal chromosomes as observed in a diploid cell are not identifiable. Such exchanges are called *translocations*. The first step in development of aneuploidy is duplication of chromosomes in a diploid cell without cell division. The result is a cell with four copies of each chromosome, or a *tetraploid* set of chromosomes.

Within the tetraploid cell some chromosomes are lost and others undergo translocations over a number of cell cycles. The chromosomal makeup of an aneuploid cell is often unstable; cells with different numbers and forms of chromosomes are continuously produced.

Cells in culture sometimes undergo only partial transformation. An example is the 3T3 cell, which was derived from a mouse embryo. These cells exhibit density-dependent inhibition, require serum growth factors, can enter the G0 state, and are anchorage-dependent, all characteristics of normal cells. However, they have the transformed characteristics of immortality and aneuploidy. Generally, for a cultured cell to become cancerous, that is, to grow as a cancer when implanted into an animal, it must be fully transformed, that is, it must possess all six of the characteristics listed in Table elsewhere in this chapter.

ONCOGENIC VIRUSES, ONOGENES, AND PROTOONCOGENES

Cancer-causing viruses are probably responsible for no more than a few percent of human cancers. However, the study of howsuch viruses work has provided major understanding of genetic changes,that occur in the induction of cancer by chemicals and radiation.

Viruses are inert particles that become active only after they have entered cells. They parasitize the metabolic machinery of cells, using ATP produced in mitochondria, using the cell's ribosomes and enzymes for protein synthesis, and so forth. Their activities are directed toward their own reproduction. Thousands of virus particles can be produced in one cell originally infected with only one or a few virus particles. Depending on the type of virus, they are released from the host cell by lysis of the cell or individually through the plasma membrane.

The minimum components of a virus are a genome of nucleic acid and a surrounding coat made of many copies of one or a few kinds of protein molecules. Many viruses are more complex, containing an additional outer membrane derived from the plasma membrane of a cell during exit from the cell. The genome of some of the viruses that invade eukaryotes is composed of double-stranded DNA; others have chromosomes composed of single-stranded DNA, or single- or doublestranded RNA. The chromosomes of oncogenic viruses are either double-stranded DNA or single-stranded RNA.

Oncogenic viruses fall into several classes. Those with chromosomes of double-stranded DNA are the papovaviruses, herpesviruses, adenoviruses, cytomegaloviruses, and hepatitis B virus. Those with singlestranded RNA chromosomes are called *retroviruses*.

Oncogenic Viruses with Double-Stranded DNA

These viruses belong to the five groups listed in Table 13-3. Most of the viruses in these groups cause cancer in animals and several are known or suspected to cause cancer in humans.

Papovaviruses

Two well studied oncogenic viruses of the papova group are the **polyoma** virus *(poly* = many; *oma* = tumors) and the simian virus 40 **(SV40).**

The polyoma virus was so named because it can cause more than ten different kinds of cancer when injected into newborn mice. It causes neither cancer nor any other disease in adult mice. Viral transformation of normal cells in culture into cancer cells was discovered with polyoma virus. There is no evidence that it causes disease or cancer in humans.

The SV40 virus was discovered in 1960 during the production of Salk vaccine against poliomyelitis virus. Polio vaccines were prepared from poliovirus grown in cultured cells derived from monkey kidneys. Monkey cells were used because of the possibility that human cells might contain undetected, harmful human viruses that would become included in the polio vaccine.

In 1960, SV40 virus was isolated from monkey cells and detected in polio vaccine produced with these cells; by this time several million people had been injected with polio vaccine containing SV40 virus. By 1962 it was clear that SV40 virus is oncogenic in rodents, causing tumors with a high efficiency when injected into baby hamsters. Fortunately, SV40 virus is not known to have caused tumors or disease in humans, and the inadvertent injection of humans with SV40 virus has had no detectable consequences. Under carefully arranged conditions, SV40 can transform human cells growing in culture.

Several other papovaviruses, known as BK, JC, and HD, have been isolated from a few human cancers, but it is not known whether they have a causative role in those cancers. Injection of JC virus causes brain cancer in monkeys.

The presence of a virus in human cancer cells does not prove that the virus has a role in causation of the cancer. Viruses are sporadically observed in normal cells, producing no discernible pathological effect. To prove that a virus causes cancer requires that injection of the virus into animals consistently induces cancer.

Adenoviruses

Adenoviruses were discovered in 1953, and in 1962 a type of adenovirus obtained from the throat of humans was found to induce cancer when injected into baby hamsters. Adenovirus infections are very common in humans, and more than a dozen of the many varieties of human adenoviruses have been found to be oncogenic in mice, rats, and hamsters.

The human adenoviruses are not known to be oncogenic in humans, but do cause transient respiratory and gastrointestinal diseases. They provide another illustration that a virus may have different effects in different species. Adenoviruses capable of causing cancer in rodents have also been obtained from cows, monkeys, and birds. More than 50 different adenoviruses, many oncogenic in animals, have been discovered.

Herpesviruses

The *herpesviruses* form a large, ubiquitous group of DNA viruses, members of which occur commonly in frogs, chickens, turkeys, guinea pigs, rabbits, monkeys, humans, and other species. In frogs, kidney cancer caused by a herpesvirus is common. *Herpesvirus saimiri*, recovered from the squirrel' monkey *Saimiri sciureus,* does not cause cancer or other ill effects in the squirrel monkey but induces a highly malignant *lymphoma* (cancer of lymph tissues) when injected into marmosets. In guinea pigs a herpesvirus causes at least one kind of leukemia. (Herpesvirus in chickens causes a highly contagious form of *leukemia*, called *Marek's disease*, which was common enough to be a serious economic problem. This leukemia is now prevented with an antiviral vaccine.

Herpesviruses are strongly implicated in several types of malignancies in humans. A kind of lymphoma (*Burkitt's lymphoma*) is common among children in certain parts of Africa. This lymphoma is almost certainly caused by a herpesvirus called the *Epstein-Barr virus (EB virus)*, named after its discoverers. It is able to transform cultured human cells, adding to the evidence that it is the causative agent in Burkitt's lymphoma.

The EB virus also causes *infectious mononucleosis*, a transient disease characterised by overproduction of lymphocytes. Why the virus can take two courses in humans is partially understood. Children in Africa who are also infected with the malarial parasite (a protozoan) have weakened immune systems. Immunological defense against EB virus is therefore probably impaired, although why this impairment leads to virus-induced lymphoma is not known. A normally functioning immune system apparently limits the tatic and bladder cancers.

Hepatitis B virus

This virus causes a severe form of hepatitis. In some individuals the hepatitis is followed by liver cancer (hepatocarcinoma). In Africa, hepatitis B virus, in conjunction with aflatoxin consumed as a contaminant of nuts and grains, is responsible for the very high rate of liver cancer in that part of the world.

Oncogenic Viruses with Single—Stranded RNA-Retroviruses

These RNA viruses, known as *retroviruses* for a reason discussed later, are of three types, B-type, Ctype, and human T-cell lymphotropic viruses.

B-Type viruses

These are also called milk viruses because they have been found in the milk of some mammals (mouse, rat, monkey, and human). The *B-type virus* was one of the first viruses proven to cause cancer in animals. It causes breast cancer in mice, but evidence that the human B-type milk virus causes human cancer is weak.

C-Type viruses

There are many *C-type viruses*, infecting a wide variety of animals. They cause leukemia, lymphoma, and sarcoma in birds and mammals. C-type viruses have occasionally been seen by electron microscopy in human leukemia and sarcoma cells, but none has been proven to cause these cancers in humans.

A few of the many extensively studied examples of C-type viruses are *mouse leukemia virus*, *Rous sarcoma virus* (originally found in chickens but capable of transforming mammalian cells), and feline *leukemia viruses (FLV)*. About 50 percent of domestic cats in the United States are infected with FLV, but only about 40 cats per 100,000 develop leukemia per year. Why so few of the infected cats develop leukemia is not known.

Feline leukemia virus is produced in the respiratory epithelium of healthy cats and is transmitted through saliva. A highly effective vaccine has been developed that prevents cat leukemia caused by this virus. There is no evidence that FLV causes disease in humans. A virus similar to FLV causes leukemia and lymphoma in cows, but it is not known whether a cat virus can infect a cow.

Human T-cell lymphotropic viruses

These viruses are so named because of their strong propensity to infect the class of human lymphocytes called T cells (see Chapter 15 for a discussion of T cells and other cells of the immune system). One of these viruses—*Human T-Cell Lymphotropic Virus I or HTLV-I*-causes adult T-cell leukemia in humans, a disease that is rare in the United States and Europe, but more common in southern Japan and other parts of the world.

Another member of this group, HTLV-III, causes *Acquired Immune Deficiency Syndrome* or AIDS. In 1986 the virus was renamed *Human Immuno-Deficiency Virus* or HIV because it infects other cells in addition to T cells. T cells have surface receptors for HIV, and it is taken in by receptor-mediated endocytosis. In T cells the virus reproduces abundantly, killing the host cell. Viruses released from a killed host cell infect other T cells, eventually severely depleting the infected individual of T cells, which are a crucial part of the immune system of defense against infection by bacteria, fungi, protozoa, and viruses. For example, individuals infected with HIV frequently die of subsequent infection by a protozoan called *Pneumocystis carinii,* which causes fatal pneumonia. Human immuno-deficiency virus also frequently invades the brain, causing death through neurological injury. HIV can infect another class of lymphocytes called *B cells* transforming them into cancer cells, giving rise to a fatal lymphoma. Further, HIV also contributes to the development of *Kaposi's sarcoma*, formerly a very rare cancer.

Human immuno-deficiency virus exhibits *latency*; that is, for an unknown reason it may infect an individual without causing any disease. Such individuals may carry the virus for several years without apparent ill effect, during which time they may spread it to other individuals. The phenomenon of latency of viruses in general is poorly understood. At first it was thought that most people who had been infected with HIV would fail to develop disease. By monitoring infected, disease-free individuals for several years, it has become apparent that most of such infected individuals eventually develop fatal disease.

Human immuno-deficiency virus undergoes frequent mutation, a property called *hypermutability*. By 1986 more than 80 different strains of HIV had been identified in the United States. Hypermutability results in frequent changes in the protein coat of the virus, the part of the virus that is recognised by the immune system. Frequent changes in the protein coat prevent the immune system from mounting an effective attack on the virus. Cold viruses (rhinoviruses) behave similarly, preventing the development of long-term immunity to colds. Hypermutability of the gene encoding the protein coat has made it impossible to develop vaccines against colds. For the same reason development of a vaccine against HIV is extremely difficult.

Some strains of HIV can apparently be transmitted only via blood, such as through blood transfusions. The virus is present in semen and is transmitted by homosexual contact. Some strains are also transmitted by heterosexual contact. HIV originated in Africa, probably in the 1950s, possibly by transmission from monkeys to humans. Adaptation of the virus to humans probably involved some mutational changes. The virus became an exploding medical problem in Africa in the late 1970s and in the United States in the early 1980s.

VIRUS INFECTION

When a virus successfully infects a cell, it multiplies, producing thousands of new virus particles in a single cell in a matter of hours. This leads to bursting of e cell (cell lysis) and release of the new virus particles, hich m in turn infect other cells. This behaviour of a virus, s called *productive* or *lytic infection*. Lytic i ction is the mechanism by which many nononcoge viruses (polio, measles, smallpox, cold, flu, etc.) reproduce in large numbers. The killing of cells in the process give rise to the symptoms of a particular viral disease in an organism, depending on which tissues are infected by the virus.

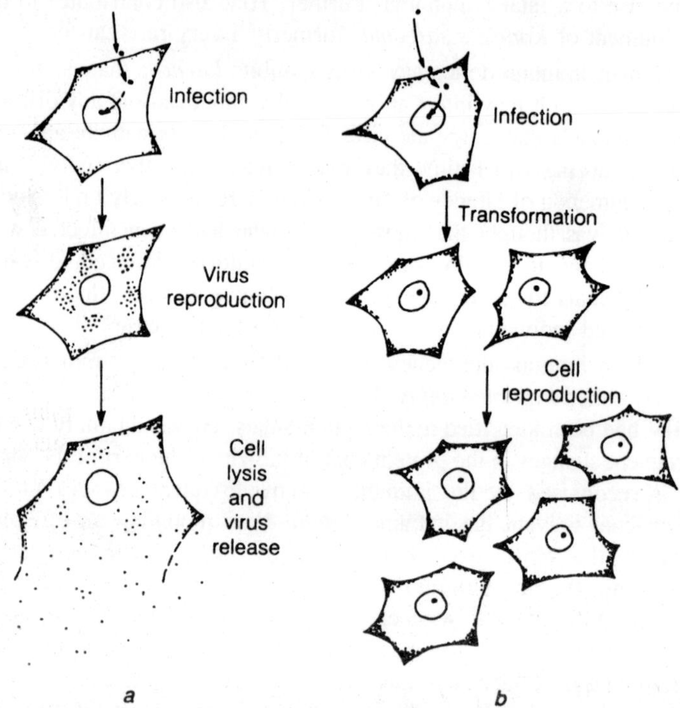

Figure 7.6: Two behaviours of viruses in cells.

Lytic infection is the only kind of infection that most animal viruses undergo. Oncogenic viruses, on the other hand, take a different route within a cell; the virus does not reproduce extensively and does not cause cell lysis. Instead of killing the cell, the virus permanently alters the cell's growth characteristics, that is, the cell is transformed by the virus, and all descendants of the cell are also transformed.

Whether or not a virus can successfully infect a cell and whether infection leads to a productive infection or transforms the cell depends on poorly understood virus-cell interrelations that are determined by the genetic properties both of the virus and the cell.

When a virus undergoes a productive infection in a particular kind of cell, the cell is said to be *permissive* for that virus. For example, monkey cells are permissive for SV40 virus but mouse cells are nonpermissive for SV40. A cell that is *nonpermissive* for a particular virus may become transformed by that virus. Thus, permissive cells are generally killed by a virus as the virus achieves productive infection, and nonpermissive cells may remain unaffected or may be transformed.

The general observation that permissive cells do not undergo transformation by a particular oncogenic virus is not a hard and fast rule; rather, permissiveness and nonpermissiveness refer to a usual pattern of virus-cell interaction. For example, mouse cells are permissive for polyoma virus, that is, *usually* they undergo lytic infection, but mouse cells may also sometimes be transformed by the polyoma virus.

Transformation by an Oncogenic DNA Virus

Much has been learned about the molecular events surrounding transformation of cultured cells by oncogenic viruses. Unfortunately, our understanding of the process is still incomplete, and we still do not know exactly how an oncogenic virus transforms a normal cell into a cancer cell.

After entry of an oncogenic virus into a transformable cell, RNA molecules that can hybridise with DNA prepared from pure virus begin to appear. This proves that genes of the viral chromosome are being transcribed. Transcription of viral genes is known to be a requirement for transformation, for example, inhibition of this transcription with a virus-specific inhibitor called *interferon* prevents the virus from transforming the cell.

A key effect of the viral gene activity is stimulation of the replication of cellular DNA. The effect may not be noticeable in cells that are already actively proliferating, but is striking when oncogenic viruses infect cells arrested in G1 by density-dependent inhibition. Within hours

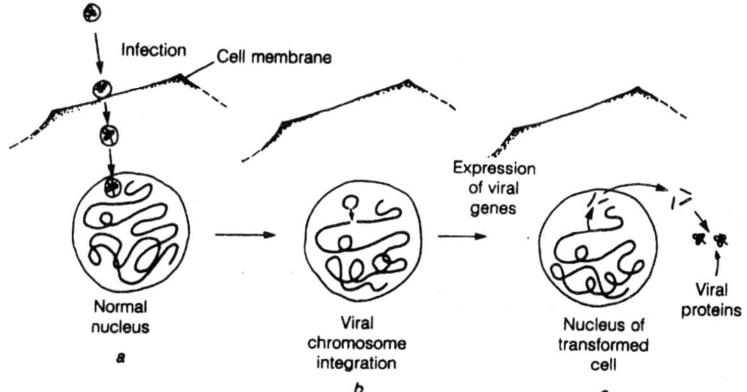

Figure 7.7 : Cell transformation by a DNA oncogenic virus.

after infection the arrested cells begin DNA replication and subsequently divide. This stimulation by the virus is important because the *fixation* of the transformed state of the cell can occur only during the S period of the cell cycle.

Once the transformed state has been induced (as soon as 24 hours after virus infection), the change is permanent, and all of the descendants of the original transformed cell are of the transformed type. Yet, no infectious virus particles can be recovered from a culture of transformed cells. (In this respect DNA and RNA oncogenic viruses differ because virus particles are usually found in and released from cells transformed by an RNA virus.) However, two kinds of experiments with cells transformed by SV40 or polyoma virus show that a complete viral genome is still present in the transformed cell.

In the first experiment two cells are fused together to form a hybrid cell. When a transformed cell (e.g., a hamster cell transformed with SV40) is fused to a permissive cell (e.g., a cell from the green monkey), virus particles begin to be produced in the nucleus contributed by the transformed cell. Subsequently, the binucleated cell bursts, releasing large numbers of infectious virus particles (in this case, SV40 particles). Since complete particles are formed, the complete viral genome must have still been present in the transformed cell.

In the second type of experiment DNA is extracted from transformed cells and added to a culture of permissive cells, a procedure called *DNA transfection*. The permissive cells take up the DNA and then begin to produce virus particles identical to the virus originally used to induce transformation, proving the presence of the viral genome in the transformed cell.

During transformation the viral genome becomes integrated into one of the chromosomes of the cell. Integration is random and can occur in any chromosome, and many viral genomes may become integrated. Integration occurs by breakage of the DNA molecule in a cellular chromosome, followed by joining of the two ends of the viral DNA to the two ends in the break of the cellular DNA. Since viral DNA molecules are closed circles, they also must break for integration to occur.

Viral transformation is an inefficient event. Even when 10^4 to 10^5 infectious virus particles *per cell* are added to a culture, only a few percent of the cells become transformed. The reason for this inefficiency is not known, but it is possible that the integration of the viral genome into a cell chromosome is the limiting step. This is suggested by the observation that conditions in which breakage of cell chromosomes is enhanced also favour a higher rate of viral transformation. Presumably, chromosome breakage provides increased opportunity for insertion of a viral chromosome. For example, cultured human cells are ordinarily transformed by SV40 with a very low frequency.

Irradiation of human cells with a sublethal amount of X rays causes breakage of chromosomes and at the same time increases manyfold the susceptibility to SV40 transformation. Also, individuals with certain genetically based disorders such as Fanconi's anemia, Down's syndrome (mongolism, caused by the presence of an extra chromosome number 21), and Klinefelter's syndrome (males with two X chromosomes and one Y chromosome) have a higher frequency of chromosome breaks.

Such individuals are not only more prone to develop cancer, but their cells grown in culture are also more susceptible to transformation by SV40. Once the viral DNA is integrated, it replicates as any other part of the cellular chromosomes and is thus inherited from cell to cell, maintaining the transformed state of a cell from one cell generation to the next. The genes carried in the chromosome of a virus code for proteins that function in the production of more virus particles.

The double-stranded DNA of SV40 codes for two proteins that form the coat of the particle and another protein, the core protein, which is complexed with DNA to form the core of the virus particle. Oncogenic viruses contain in addition one or two genes that are responsible for transformation of their host cells.

The SV40 virus contains two genes, designated t and T, that are essential for transformation. Adenoviruses carry a gene called *EIA* that is responsible for transformation. Indeed, cell transformation can be achieved by integration into a host cell chromosome of only the fragment of the viral chromosome that carries the transformation gene or genes but has none of the genes that function in viral reproduction.

Transformation by RNA Viruses

In general, RNA oncogenic viruses transform cells by the same principles as DNA oncogenic viruses. However, for transformation to occur, the single-stranded RNA chromosome must first be copied into a doublestranded DNA molecule. The key to this process is the enzyme reverse transcriptase, the gene for which is carried by all oncogenic RNA viruses. After infection of a susceptible cell, reverse transcriptase catalyzes the polymerisation of a DNA chain using the singlestranded viral RNA as a guiding template. This creates a hybrid double helix, one strand of which is the original viral RNA molecule and the other

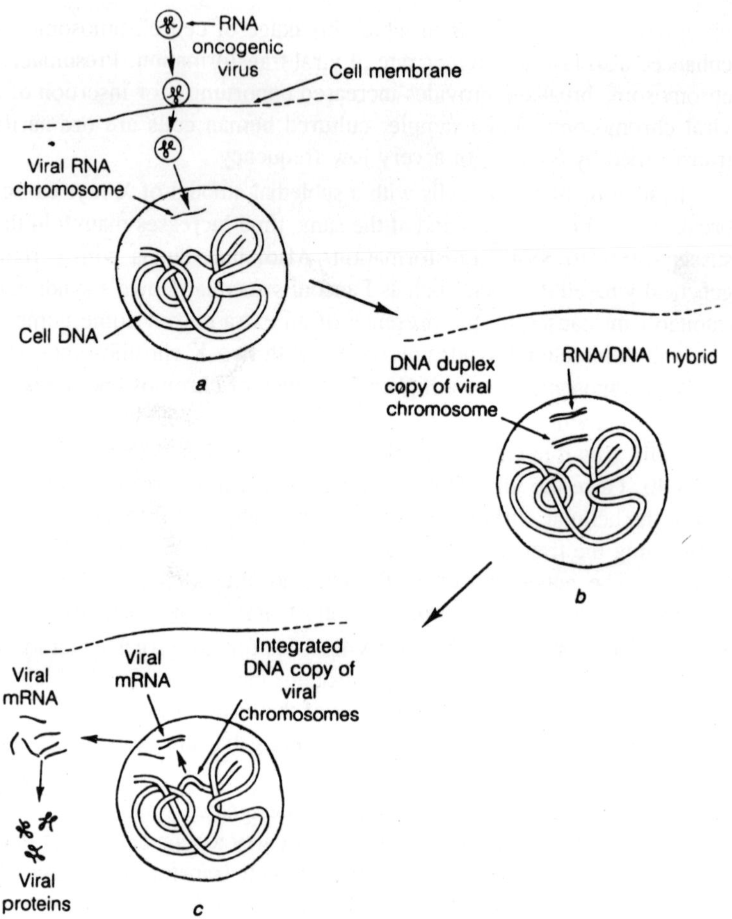

Figure 7.8: Transformation by a retrovirus.

strand of which is the newly synthesized DNA chain. The RNA is digested from the double helix and a DNA/DNA duplex is formed. The overall effect is to transfer the genetic information of the single-stranded viral RNA into a DNA duplex, the precise reverse of transcription. The DNA duplex is then integrated into a cellular chromosome.

The reverse transcription of RNA into a DNA version as a part of transformation is the basis of the name retrovirus (retro = backward). The integrated DNA may be repeatedly transcribed in its entirety to yield single-stranded RNA molecules, which may become packaged into virus particles, which are released singly from the transformed cell without lysing the cell. Oncogenic RNA virus particles acquire an

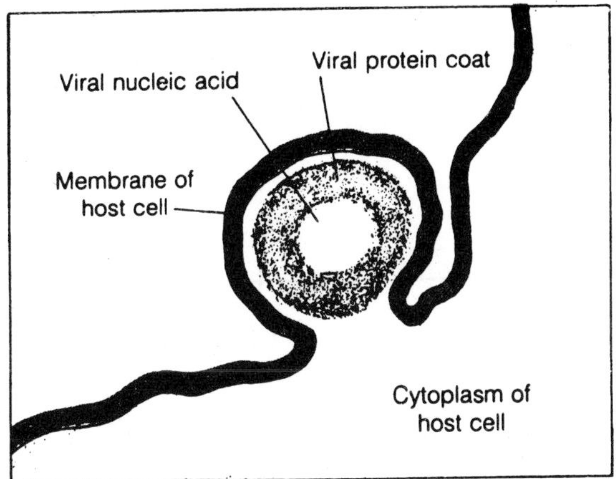

Figure 7.9: A drawing of the structures in the micrograph. This particular virus causes leukemia in animals.

additional coat derived from the plasma membrane as they leave the cell by a budding process.

Oncogenes

As in the case of DNA oncogenic viruses, the integrated DNA version of viral RNA contains at least one gene whose activity converts the cell to the transformed state.

The chromosome of most retroviruses carries only a few genes. Many conform to the pattern shown for Rous sarcoma virus (RSV) in Figure elsewhere in this book. The *gag* gene codes an internal protein that complexes with the RNA to form the core of the virus. The ***env*** gene encodes the coat protein subunits. The ***pol*** gene codes for reverse transcriptase, and the *one* gene encodes the protein responsible for cell transformation. The *one* gene in RSV is given the more specific name src, for sarcoma.

The *src* gene was the first oncogene in a retrovirus to be identified and studied intensively. Its cancer-causing property was proven with a temperature-sensitive mutation. Cultured chicken cells infected with a virus containing a mutation in the *src* gene remained normal at 39°C (protein encoded by the gene is not functional) but immediately became transformed when cultured at 33°C (protein encoded by the gene is functional). The mutated virus reproduced normally in permissive cells at both temperatures, which showed that the *src* gene product was unnecessary for virus reproduction.

Reverse transcriptase has been used to convert the RNA of a

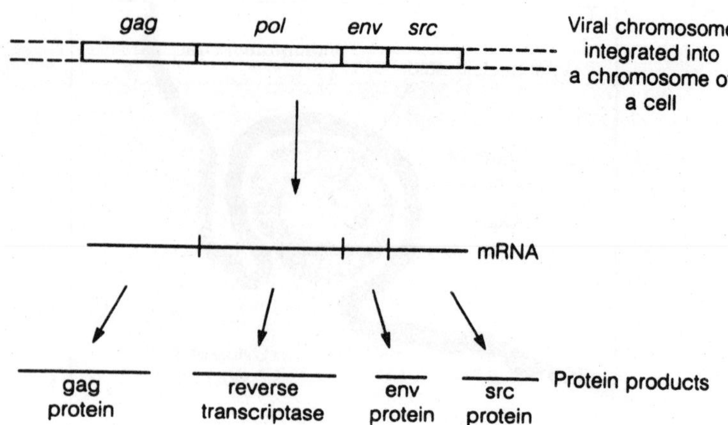

Figure 7.10: The chromosome of Rous sarcoma virus in its DNA duplex form integrated into a cellular chromosome. The DNA is transcribed into a single mRNA molecule that is translated into four proteins.

nonmutated Rous sarcoma virus into a double-stranded DNA copy (cDNA) in a cell-free system. A fragment of the cDNA containing the *src.* gene has been prepared and shown to be sufficient by itself to transform a cell.

More than 25 different oncogenes have been identified in retroviruses that infect birds and mammals. Some of these, including the *src* gene, encode a *protein kinase*. The kinases encoded by these oncogenes are found in the cytoplasm bound to the inner side of the plasma membrane. These kinases bring about transformation of a cell by phosphorylating particular proteins, but little is known about these proteins or how their phosphorylation results in such dramatic changes in the cell. Other oncogenes encode proteins that are localised in the nucleus, but no enzyme activity has been identified for these as yet. Still others encode growth factors. Other kinds of oncogenes encode proteins that disrupt regulation of cell behaviour by affecting some other part of the mechanisms that regulate cell behaviour.

A normal cell can also be transformed to a cancer cell with carcinogenic chemicals or by radiation. Since most of these agents are proven mutagens, they presumably transform cells by mutating one or another gene involved in regulating cell behaviour.

At least some of the genes that are altered by chemical carcinogens and radiation in a transformed cell are closely related to the oncogenes of viruses, as shown by the following kind of experiment. Total DNA was isolated from human nonvirally induced cancer cells. One or more altered genes contained in that DNA were presumed to be responsible for the cancerous state of the cells. The presumption was tested by a *transfection assay* in which DNA purified from cancer cells was added to a confluent monolayer of mouse 3T3 cells in culture under conditions that promoted the uptake of the DNA by the cells.

Mouse 3T3 cells are normal in that they show density-dependent inhibition, anchorage dependence, a high serum requirement, and ability to arrest in G1 and to enter G0. They have only two transformed characteristics-immortality and aneuploidy. A few of the millions of 3T3 cells in the densityarrested, confluent monolayer became fully transformed by uptake of the human cancer DNA. These cells were evident because they were released from density-dependent inhibition (arrest in G 1) and resumed proliferation, forming foci of piled up growing cells.

In the transfection experiment with DNA from a human cancer it was proven that the transformed 3T3 cells had taken up human DNA because they now contained human Alu sequences. The Alu sequence is repeated in over 900,000 copies dispersed throughout human DNA, but the sequence is absent in mouse DNA.

The transformed 3T3 cells had taken up other human DNA sequences in addition to the one causing transformation, making it impossible to identify which particular sequence was responsible for transformation (and the cancer). Therefore, DNA was isolated from the transformed 3T3 cells and added to another confluent culture of normal 3T3 cells. Because the added DNA was mostly mouse DNA with a very limited number of human DNA sequences, it could be reasonably expected that transformed cells now appearing would contain only that human DNA sequence or sequences necessary for transformation.

Such proved to be the case. Only one human sequence was needed to transform mouse 3T3 cells, and moreover this sequence contained a copy of the sequence of the oncogene known as *ras* present in a virus that causes sarcoma in mice (murine sarcoma virus). Hybridisation experiments have subsequently shown that all *normal* human and other mammalian cells contain a gene corresponding to the *ras* gene. However, the *ras* gene in the human cancer cells used in the above experiment has a mutation in one codon, changing a single amino acid in the protein encoded by the gene. Thus, mutation in the normal *ras* gene altered the function of the protein it encodes so as to convert a normal cell to a cancer celi.

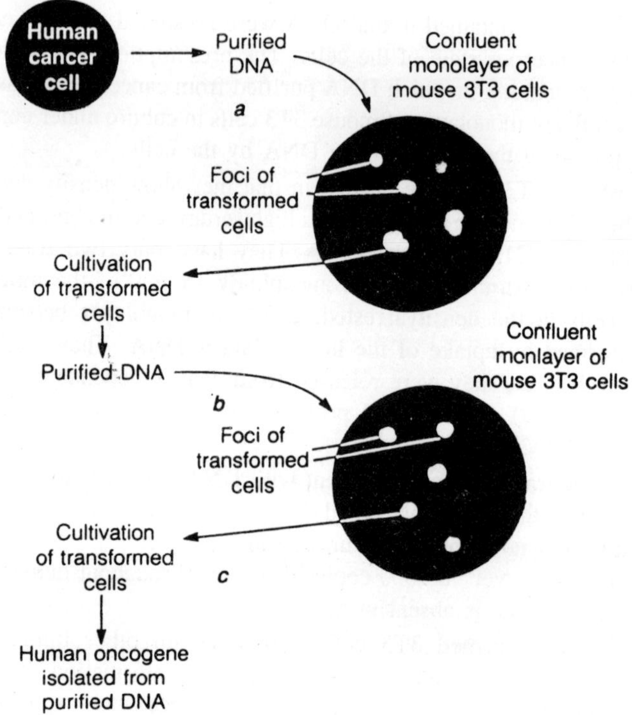

Figure 7.11: Identification of a human oncogene by DNA transfection.

Subsequently, over 20 other genes have been found in normal mammalian cells that correspond to oncogenes present in oncogenic viruses. Some of these genes have been found in a wide variety of unicellular and multicellular eukaryotes, such as, yeast and *Drosophila.* Often, small differences between the cellular versions and the viral versions are responsible for the oncogenic nature of the viral version. The genes in normal cells are called *protooncogenes*.

These genes are believed to be important in regulation of cell reproduction and differentiation. The right kind of mutation in a protooncogene converts it to an oncogene. Such cellular oncogenes, found in cancer cells are called *c-onco-genes* or *c-oncs* to distinguish them from oncogenes, called *v-oncs*, carried by oncogenic viruses.

Many kinds of human cancers have been found to contain c-oncogenes derived by alteration of a protooncogene. In addition, using the DNA transfection assay described above in the original identification of the cellular *ras* oncogene *(c-ras)*, several c-oncs have been found in human cancers that have no known viral counterpart.

In the transfection assay used to identify the *ras* gene the test cell was the 3T3. These cells are already transformed in two characteristics: they are immortal and aneuploid. A repeat of the transfection experiment with cancer cell DNA using normal test cells (diploid and lacking immortality) freshly explanted from mouse tissue failed to yield transformed cells. However, when two different oncogenes were introduced into normal cells, full transformation was achieved. Presumably each oncogene is responsible for a different part of transformation.

Protooncogenes are converted to cellular oncogenes in at least two ways: the first is mutation, as already cited. The mutation must occur at specific sites to alter the function of the gene product in the proper way. In the case of the *ras* protooncogene the mutation must change the amino acid either in the 12th or the 61st position of the polypeptide product. These two particular amino acid positions are important because they change the function of the protein encoded by the *ras* gene in a specific way.

The second kind of change in conversion of a protooncogene to an oncogene is one that results in overexpression of the gene without change in the encoded protein product. This can happen by any of the following: a mutation in the promoter that increases its effectiveness; movement of the protooncogene to a new location, where it comes under the influence of a strong promoter; or an amplification in the number of copies of the gene.

Movement of genes occurs by breakage and rejoining of chromosomes in which two chromosomes exchange segments, a process called chromosome translocation, which can be induced by radiation and some kinds of chemical mutagens. For example, the protooncogene *abl* (present as the v-onc in a mouse leukemia virus) has been moved from its normal location in chromosome nine to chromosome 22 in almost all cases of chronic myelogenous leukemia in humans. In its new position it is expressed at a much higher rate. The protooncogene *myc* (present as a v-onc in avian myelocytomatosis virus) is amplified in some human cancers.

Viral oncogenes in retroviruses were almost certainly derived from mRNA copies of protooncogenes in the distant past. In a rare molecular accident an mRNA molecule of a protooncogene probably became incorporated into the RNA chromosome of a nononcogenic virus. Part of the support for this idea is the presence of introns in cellular protooncogenes and the absence of all such introns in the corresponding v-oncs of viruses.

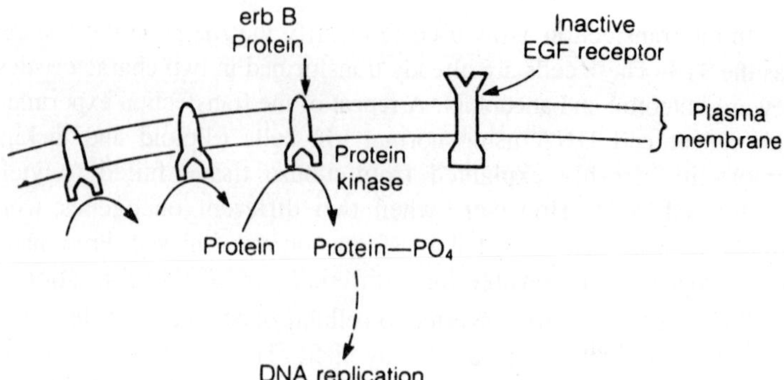

Figure 7.12: The erb B protein encoded by an oncogene in avian erythroblastosis virus is homologous to the protein kinase domain of the EGF receptor. The constitutive protein kinase activity of erb B protein results in cell transformation.

Several of the protooncogenes present in mammals have been found in a wide variety of eukaryotes, including *Drosophila,* protozoa, and yeast cells. This indicates that these protooncogenes are ancient genes and probably encode proteins that are fundamental to regulation of cell reproduction in all eukaryotes. The gene in yeast known as *CDC-28,* which encodes a protein that functions in the Start switch in G1, has a high degree of homology of nucleotide sequence with the src gene of mammals. This homology indicates that the *src* gene evolved from an ancient gene important in regulating initiation of DNA replication.

Growth Factors and Oncogenes

How conversion of protooncogenes to oncogenes disrupts normal cell behaviour is not known, but clearly at least some protooncogenes encode proteins that are part of the mechanism by which growth factors regulate cell reproduction. For example, the receptor in the plasma membrane for epidermal growth factor (EGF receptor) is encoded by a gene that shows extensive homology with *v-erb B,* one of two oncogenes in avian erythroblastosis virus.

The product of v*erb B* lacks the EGF binding domain but does have the domain that acts as a protein kinase. Without the EGF binding domain to act as a modulator, the protein kinase domain is constantly expressed. This is equivalent to constant stimulation of a normal cell by EGF (although no EGF is present) and consequent loss of regulation of cell reproduction by EGF.

The oncogene *sis,* present in simian sarcoma virus, is homologous to the normal cellular gene encoding platelet-derived growth factor (PDGF). Transformation with simian sarcoma virus results in a constant high level of synthesis of PDGF, which obliterates regulation of cell reproduction by extracellular PDGF.

Thus, recent research has tied together viral oncogenes, cellular protooncogenes, and mechanisms of regulation of animal cell reproduction by growth factors. Discoveries in each of these areas directly impinges on understanding of the other two, producing a cooperative effect that will lead to a molecular understanding of regulation of cell reproduction and the loss of regulated behaviour in a cancer cell.

8

AGEING

The terms *ageing* and *senescence* are familiar, but unfortunately they are rather imprecise. There is no general agreement among gerontologists about how they should be used. Usually *senescent* and *senescence* are used when talking about the changes which occur during the period of obvious functional decline in the later years of an animal's life-span. Some people use the term *ageing* for the same processes and period. Others use it in a much more general way, with *ageing* meaning simply growing older, and *ageing changes* being any changes related to age, regardless of when in the life-span they occur.

Thus the onset of puberty might be described as an ageing change, but not as a senescence change. Although there may be advantages in using *ageing* in this general way, it is rather difficult to do so, because in common language *ageing* implies something more than simply getting older. For example, it would be unusual to talk of an ageing child.

We would normally refer to a developing child, because in everyday English the word *ageing* carries with it the idea of decline and deterioration. Most biologists have tended to accept these connotations, and think of ageing as occurring only after maturity has been reached. In fact, as can be seen from some of the definitions given below, the terms *ageing* and *senescence* are frequently used interchangeably.

The have been many formal definitions of ageing processes and senescence. Medawar (1952) suggested that senescence could be defined as during ageing are deleterious; they increase the chances that an animal will die. Ageing therefore involves a decrease in the ability of an animal to cope with its environment. Secondly, the deleterious age-related changes are cumulative. Death, the ultimate result of ageing, is sudden, but the process of ageing involves a progressive increase in the probability of dying. A third characteristic of ageing and senescence, which is implicit in most of the definitions which have been given, is that the processes involved are common to all members of a species and are inescapable consequences of getting older. That is to say, ageing and senescence are fundamental intrinsic properties of living organisms.

AGEING AND THE LENGTH OF LIFE

Sometimes it is helpful to visualise the characteristics of ageing and the relation between ageing processes and the length of life in the way illustrated. On the left of the diagram the ageing of an individual has been shown as a loss of *vitality* with time. *Vitality* is a term which has been used by a number of gerontologists, including Raymond Pearl, one of the founders of modern experimental gerontology. By *vitality* we mean nothing more than the ability to sustain life, the ability to survive.

This gets less as age increases. Quite deliberately no scale has been put on either of the axes. If the time scale was linear, it would imply that ageing is a linear function of time; there is no reason to think that this is true. By omitting a scale we are leaving open the question of whether the rate of ageing increases, decreases, remains constant or fluctuates in any particular time interval. The straight line parallel to the time axis is a death threshold: when vitality falls below this threshold, death occurs.

The level of the death threshold reflects the severity of the conditions to which the individual is exposed. If conditions are severe, the threshold will be higher, and hence the length of life will be shorter.

The right-hand section is an attempt to make the picture ageing and longevity a little more realistic. Ageing and times death for four individuals are shown. The first point to note that the individuals start life with different amounts of vitality. Secondly, the rate of loss of vitality, or rate of ageing, is not the same for all individuals.

The justification for these two features will be discussed in later chapters For our present purpose it is sufficient to say that there is evidence that longevity is influenced by an animal's genotype and, since there are usually genetic differences between individuals in a population, either the rate of ageing or the initial level of vitality or both of these characters may be different in different individuals.

The representation of the death threshold has also been modified in the right-hand part of the figure. The environmental challenges that an animal has to cope with are never completely constant, so the death threshold has been shown as fluctuating around a mean value.

Simple two-dimensional illustrations such as those provided may sometimes be misleading when they are used to visualise a little-understood but undoubtedly complex process such as ageing. It is not difficult to criticise some of the features of the diagram. For example, even in the right-hand part of the figure it has been assumed that there is no relation between the rate of ageing and the death threshold. This is not necessarily true.

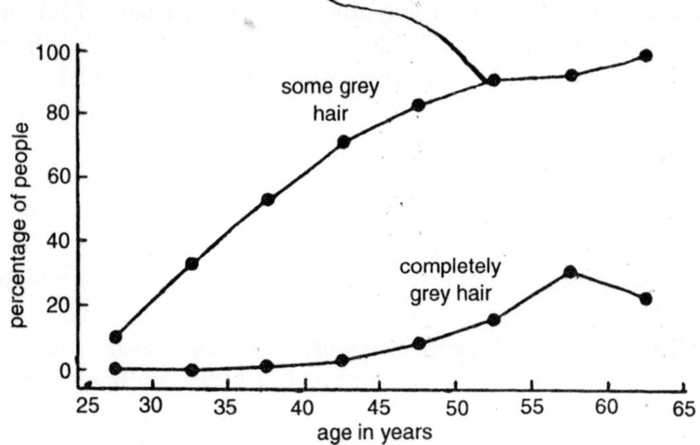

Figure 8.1 : Rate of hair greying for 3872 men and women with medium colour hair.

Suppose that the cause of the increase in the death threshold is a disease epidemic or a very cold spell. Is it correct to assume, as is implied in the diagram, that the rate of ageing is unaffected by these factors? There is no justification for this assumption, because we do not know the extent to which the rate of ageing depends on environmental stresses. Similarly, the diagrm shows only one death threshold and therefore suggests that the factors which may cause death are, at any instant, identical for all individuals. This is most unrealistic.

In spite of these and other defects, the illustration is of value because it does make it easier to see how complex the relation between ageing processes and the time of death is likely to be. Individual A ages more rapidly them individual C, yet it has a longer life-span than C. Individual D lives longer than B, yet if these individuals had been challenged by a very severe stress at the beginning of their lives, B, the individual with the shorter life-span, would have had a greater chance of surviving than D. The point to note is that the length of life does not depend simply on the rate at which ageing processes are occurring.

AGEING CHANGES IN MAN

One of the most obvious signs of human ageing is the greying of hair. With increasing age more and more people have some grey hair. This figure is based on data obtained from a survey of Australian men and women, in which hair-greying was assessed in three categories : *complete greying, any greying,* and *no greying.*

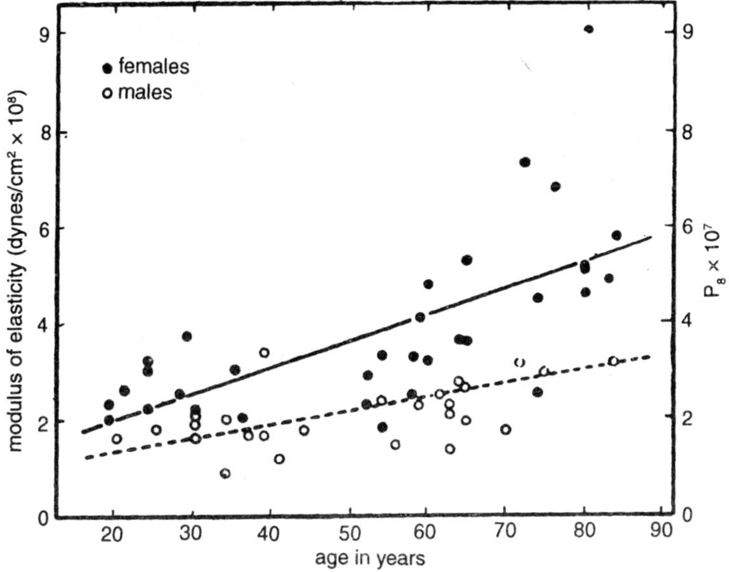

Figure 8.2 : Age-related changes in the in vivo elasticity of human fore-arm skin.

Although there were problems in distinguishing grey hair in fair-haired people, when allowance was made for this, there was no evidence of any differences in the rate of greying among people of different hair colours, or between men and women. It should be noted that greying begins before the age of 30 and steadily increases. The way in which skin elasticity changes with age. This measurable change in skin properties is probably associated with one of the obvious visible signs of ageing, namely changes in skin texture and appearance.

The figure shows that in both sexes there is a gradual rise in the modulus of elasticity, but that this "stiffening up" is more marked in women than in men, in spite of the fact that women have the longer life-span. As we shall discuss, the age related change in skin elasticity is probably caused in part by changes inth collagen molecules which form a high proportion of the protein in skin.

Apart from visible changes in appearance, one of the most familiar signs of ageing is a general loss of strength. Several different methods have been used to measure this change. The results from measuring hand-grip strength. The values plotted are the averages for each decade of life, and they show that after an initial rise there is a decline in grip strength for both men and women.

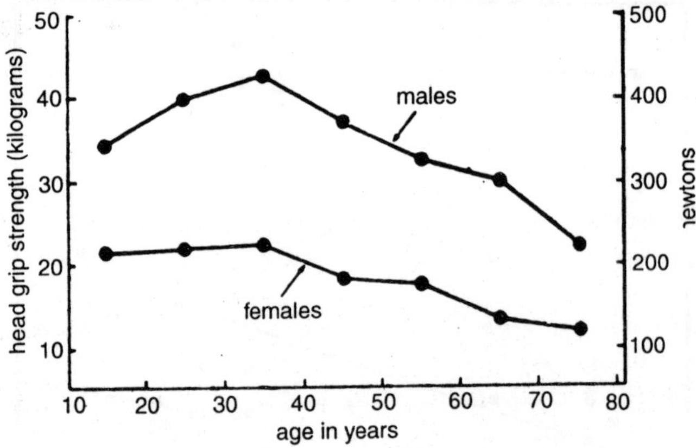

Figure 8.3 : Changes in hand grip strength with age.

These data come from a preliminary study of a number of physiological functions in over 400' people in Hiroshima. The aim of the study is to measure a number of variables which are known to change with age, and from these to attempt to determine "biological age" and ultimately to relate this to mortality in the subjects surveyed. The results from two other series of tests made in this survey; both blood pressure and the vital capacity of the lungs show marked changes after the age of 40. Hearing, visual acuity, vibration perception, manual dexterity and mental reaction time were also tested, and all showed deteriorations, particularly beyond the age of forty.

Some of the most striking age-related changes are revealed by studying the responses of the body to a physiological challenge. The pH, ion concentration, temperature, levels of metabolites and of food substances in the internal environment of mammals are normally maintained within very narrow limits. Numerous feed-back mechanisms tend to restore the composition of the internal environment to normal if a natural or artificially applied stress causes it to change.

It is clear, however, that these homeostatic control mechanisms

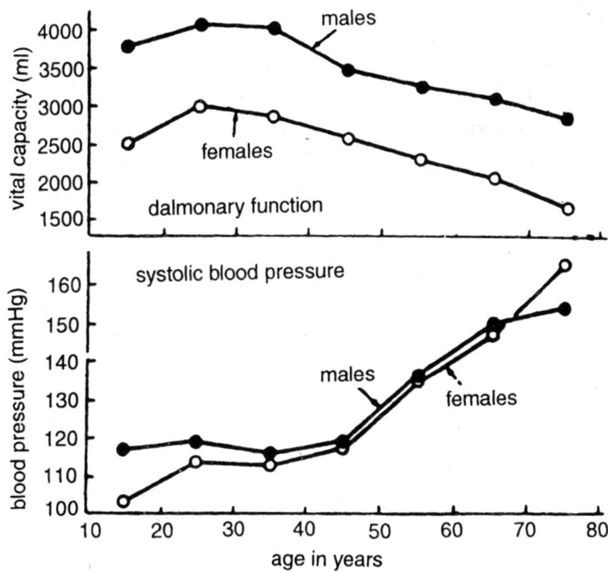

Figure 8.4 : Changes in pulmonary function and blood pressure with age.

become less efficient as age increases. For example, normally the glucose level in blood shows very little change with age, but if glucose is injected into the blood, the rate of restoration to the normal level shows a marked dependence on age: it is far slower in old people than in young.

This decline in the efficiency and effectiveness of the homeostatic regulatory mechanisms could be due to deficiencies in one or all of a number of interdependent tissues and cells. In the case of glucose, the muscles and liver are important for its storage, the pancreas is involved in the production of insulin, which regulates storage and mobilisation, and the kidneys are involved in getting rid of excess amounts.

Nathan Shock has made extensive studies of ageing changes in man and other mammals, and has concluded that the functions which show the greatest change are those which involve the co-ordinated activity of a number of organ systems.

Those which show little or no change usually involve only a single organ or system. For example, nerve conduction velocity changes by only about 10% between the ages of 20 and 90, whereas the maximum breathing capacity decreases by 50%. The former is a measurement of the performance of nerves only, the latter depends on the efficiency and co-ordination of both the nervous and muscular systems. Shock (1974) has suggested that from a physiological point of view, ageing involves primarily an impairment or breakdown of control mechanisms. The more complex a task is in biological terms, the more likely it is to show an age-related deterioration.

It is impossible to consider all the age-related changes in physiological functions that are known to occur in man. Others will be referred to in later chapters. For the moment it is sufficient to say that almost every system which has been studied shows some change. There is no difficulty in finding measurable age-related changes.

The problem is deciding what these changes mean in relation to "biological age" and the probability of dying. One of the best indicators of age is hair-greying; yet it is difficult to believe that this has much to do with the increased likelihood of dying as age increases. Skin elasticity is a good indicator of age; yet, according to the results, if judged only by skin elasticity, a woman of 30 has the same biological age as a man of 80.

In the study of Japanese subjects, many of the tests gave results which showed that the main changes occurred in the latter half of life, but hair-greying and skin elasticity changes certainly begin long before this. Which of the age-dependent changes is going to provide the most useful measurement of the rate of ageing? At present it seems unlikely that any one measurement will ever be sufficient.

A "testbattery", as Alex Comfort calls it, is needed. Comfort has suggested that it is now possible and desirable to organise a battery of tests to measure the rate of ageing in normal human subjects. He argues that if enough suitable characteristics are measured, then it should be possible to estimate biological age and the rate at which ageing changes are progressing.

It is important that we should be able to assess biological age. Our understanding of ageing may soon progress to the point where it is possible to suggest which factors do or could influence the rate of ageing in man. Unless we have some measurement of ageing other than mortality rates, it is going to take an unrealistically long time to see whether a drug (or environmental or nutritional factor) is capable of affecting the rate of ageing processes.

AGEING, INJURY AND DISEASE

One of the signs of ageing which is all too familiar to many of us is the increasing incidence of many different types of disease. As we get older, we frequently begin to show symptoms of arthritis, osteoporosis, atherosclerosis, or malignant tumours, for example. When old people die, it is often very difficult to pinpoint the exact cause of death, because pathological changes can be seen in so many tissues and organs. In developed countries, the major causes of death are at present cardiovascular diseases and cancer, but it has been calculated that if it was possible to eliminate both of these, average lifespan would be increased by only 6 or 7 years. If we did not die of cancer or cardiovascular disease, we would soon die of something else.

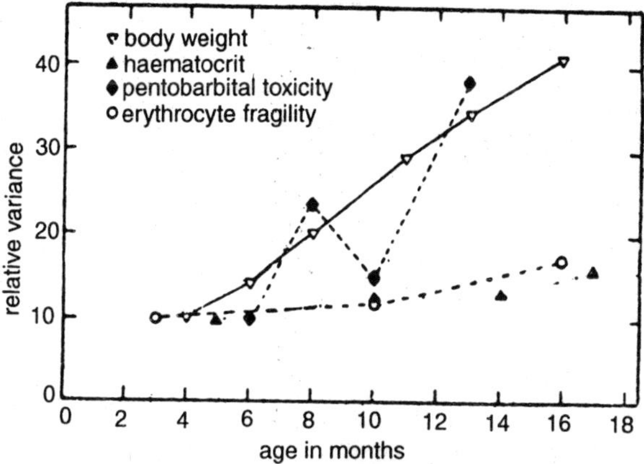

Figure 8.5 : Age-related changes in the variance of body weight, erythrocyte fragility, pentobarbital toxicity, and haematocrit for male mice.

Most of the diseases which we associate with old age may well be initiated long before they become clinically obvious. For example, the early signs of changes in the arteries which are thought to precede atherosclerosis-a condition which may ultimately cause death by a stroke or heart attack-can often be detected *post mortem* before the age of 20. With increasing age, the number and size of the lesions in the arteries increase, and their structure changes. The problem is, are diseases like this the causes or the consequences of ageing processes? Are they innate and inevitable, or are they caused by environmental factors?

From time to time it has been suggested that age-dependent loss of vigour is simply the result of pathological initiated by earlier disease or by other forms of wear and tear and injury to the body. In the early part of this century, Metchnikoff, the discoverer of phagocytosis, proposed that many of the adverse effects of increasing age would be prevented if harmful microbes could be eliminated from the body.

In particular, he believed that the bacteria in the large intestine Aere harmful because they produced toxins which passed into the lymph and blood, and led to degeneration of other parts of the body. He suggested that elimination of the harmful gut bacteria, elimination of syphilis, and suppression of alcoholism would prevent many of the pathological symptoms of old age and probably increase the length of life. Although in this extreme from the idea that ageing is due to disease now receives little support, it cannot be denied that previous disease and injury must make some contribution to the age-related increase in the probability of dying.

Animals are constantly exposed to infections which are likely to cause minor damage to cells and tissues. In mammals, teeth are damaged and lost, and cannot be replaced. Injuries such as limb fractures occur; usually they heal, but the healed limb may not be as efficient as it was before the fracture occurred. Thus, as an animal gets older it will-simply because it has lived longer-accumulate more and more of the type of damage associated with minor diseases and injuries.

At present, few people seem prepared to accept that this accumulation of damage due to earlier infection and injury of environmental origin is the sole cause of senescence. Rather, senescence is seen as an innate process which would occur even in the absence of injury and disease. Separation of the environmental and innate components of age-related deterioration is difficult, but several pieces of evidence suggest that the environmental component is not as great as might be expected.

First of all, there is no evidence to suggest that the maximum human life-span in modern industrial societies is any greater than that in underdeveloped countries, or than that in previous centuries. If previous disease history is important, we might expect that the antibiotics and immunisation procedures now used would reduce the damage due to disease and thus increase life-span. Average life-spans are now greater in western societies, but the maximum longevity seems not to have been changed.

A second line of evidence suggesting that disease may not be an important cause of ageing is that prodided by studies of germ-free mice and rats. The results obtained by different workers are not entirely consistent, but in general they suggest that, although the average age at death may be different, the maximum life-span is not increased in a pathogen-free environment.

Curtis and his colleagues have attempted to test experimentally the idea that ageing is due to residual damage from disease and injury. They subjected mice to repeated doses of substances such as nitrogen mustard or tetanus toxin over a period of many months. When the stress was removed, they found that the animals lived as long as those which had never been treated. Thus these stresses did not appear to cause an increase in the rate of ageing. Comparable results have been obtained with *D. melanogacter* subjected to thermal stresses. However, we know that some agents, in particular sub-lethal doses of ionizing radiation, undoubtedly do affect longevity adversely.

All that we can conclude from the evidence outlined above is that there is no compelling reason to believe that ageing changes are caused solely by the cumulative residual effects of disease, injury, and wear and tear. Environmental factors such as disease and injury must play some part, but it is not clear how big a part it is.

THE LENGTH OF LIFE

In spite of the difficulty of obtaining adequate life-table data for most animals, enough is known about the longevity of the various species to make it quite clear that one of the most important facts which any generalised theory of ageing must be capable of explaining is that different species have different characteristic life-spans. In this chapter we shall survey briefly some of the data which show the wide range of life-spans found in different groups of animals, and then go on to consider the extent to which the differences in longevity found within a species are due to hereditary factors.

The Life-span of Mammals

Table elsewhere in this chapter gives the maximum recorded life-span of a number of common mammals. The data are taken from the *Biology Data Book (1972)* which gives a far more extensive list than that given in the table, and also gives the source of the longevity record for each species. Before discussing the data it is necessary to consider how representative of the true longevity potential of each species the figures are likely to be.

The life-span given is the maximum length of life recorded for animals kept in captivity in zoos or laboratories, or as farm animals or domestic pets. This means that many of the figures are based on very small samples. The longevities of many thousands of laboratory mice and domestic cats have been recorded, so the figures given are probably a reasonably good indication of the maximum life-spans of these species.

On the other hand, the numbers of brown bears or porcupines which have been kept in captivity are relatively small, and the figures given may not be at all representative of the maximum longevity attainable. The more animals that have been kept in captivity, the greater the maximum recorded longevity is likely to be, simply because large samples are more likely to include a potentially long-lived individual than small samples. There is another reason why the longevity records

Table 8.1 : Maximum recorded life-spans for mammals kept in captivity

Species	*Common name*	*Maximum life-span*
Order Artiodactyla		
Capra hircus	wild goat	18 y
Camelus bactrianus	Bactrian camel	29 y 5 mo
Hippopotamus amphibius	hippopotamus	51 y
Order Perissodactyla		
Tapirus terrestris	South American tapir	30y 5 mo
Equus caballus	domestic horse	46 y
Rhinoceros unicornis	Indian rhinoceros	40 y
Order Proboscidea		
Elephas maximus	Indian elephant	70 y
Order Tubulidentata		
Orycteropus afer	aardvark	12 y
Order Pinnipedia		
Phoca vitulina	harbour or common seal	>34 y
Order Carnivora		
Cants familiaris	dog	20 y
Felis catus	domestic cat	28 y
Ursus arctos	brown bear	36 y 10 mo
Mustela vison	American mink	10 y
Order Rodentia		
Cavia porcellus	guinea pig	7 y 6 mo
Hystrix brachyura	porcupine	27 y 3 mo
Mus musculus	house mouse	3 y 6 mo
Rattus rattus	black rat	4 y 8 mo
Order Lagomorpha		
Oryctolagus cuniculus	European rabbit	13 y
Order Edentate		
Dasypus novemcinctus	nine-banded armadillo	10 y 4 mo
Order Primates		
Galogo crassicaudatus	thick-tailed bush baby	14y
Cebus capucinus	white-throated capuchin	>40 y

Table Contd.

Species	***Common name***	***Maximum life span***
monkey		
Gorilla gorilla	gorilla	>39 y 4 mo
Pan troglodytes	chimpanzee	>44 y 6 mo.
Pongo pygmaeus	orang-utan	>*50* y
Order Chiroptera		
Eidolon helvum	African fruit bat	21 y 10 MO
Order Insectivora		
Erinaceus europaeus	hedgehog	4 y 2 mo
Sorex palustris	Northern water shrew	I y 6 mo
Order Marsupialia		
Macropus robustus	wallaroo or euro	19 y 7 mo
Order Monotremata		
Tachyglossus aculeatus	Australian echidna	49 y 5 mo
Ornithorhynchus anatinus	platypus	17 y

y=years mo=months

of animals which are not commonly kept in captivity may be relatively low; the more individuals of a species that are kept in captivity, the more adept human beings become at keeping them. With experience, we understand the species environmental requirements more thoroughly, and are likely to be able to cure minor illnesses. Thus the figures which are given in the table are probably only a very crude indication of the maximum longevity for many species. Nevertheless, they do show how very different the life-span of different species can be.

Since there is so much variation in the life-spans of mammals, it is natural to look for reasons for the differences between species. Is there any structural or functional characteristic of the long-lived species which is not shared by the short-lived species? One of the things which is apparent from the table is that in general the larger mammals tend to have the longer life-spans.

The possible significance of this relationship was realised by Rubner in 1908. He found that for five species of domestic mammals with very different life-spans and body weights, the total basal energy expenditure during the life-span was approximately constant at about 200 kilocalories per gram of body weight (840 J/kg).

In other words, high metabolic rate was associated with short life-span and *vice versa.* The relation between size and longevity is brought about by the fact that there is a strong correlation between basal metabolic rate and size-small animals have high metabolic rates. Rubner found that although his hypothesis (that longevity was determined by metabolic rate, which in turn depended on size) seemed to hold for the five species of domestic animals for which he had data, man did not fit the general pattern.

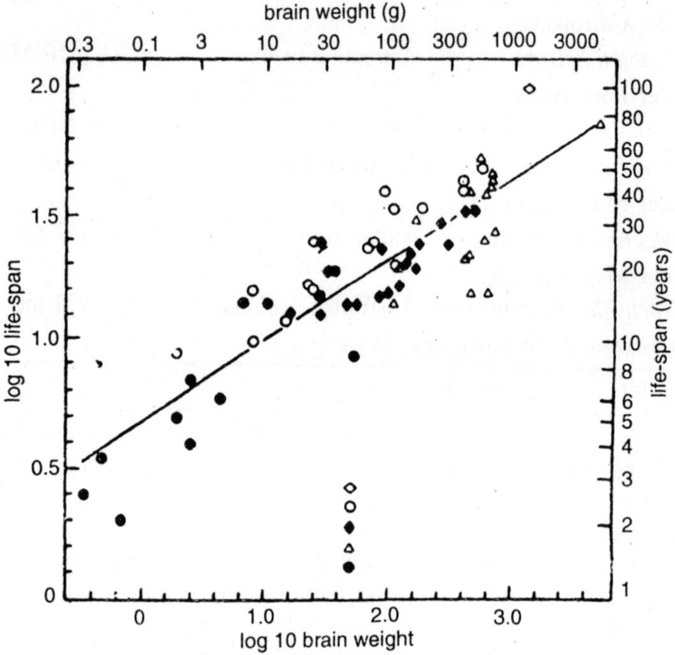

Figure 8.6: Relation of life-span to brain weight for different species of mammals.

Man's life-time energy expenditure was three times higher than that of other animals; he seemed to live too long for his size. The relation between body size and longevity was re-examiny Sacher (1959) using data for 63 species of terrestrial placental mammals. Sacher's data for both body weight and maximum recorded life-span are plotted on a logarithmic scale. It can be seen that there is a clear correlation between the two variables.

However, there are certain peculiarities apparent. First of all, as with Rubner's data, man is an exception. The value for the life-span of man seems to be more than three times greater than would be expected from his body weight; on the basis of the regression line, man should have a life-span of only about 30 years. Secondly, there is a marked tendency for all the primates to have lifespans which are too long (the points cluster above the line), and for the rodents to have life-spans which are too short (the points cluster below the line).

Since there is a good but imperfect correlation between body weight and longevity, the next question which was asked was whether a better correlation would be obtained if the weight of some part of the body rather than total body weight was compared with longevity. Most organs of the body show a close correlation with total body weight, but one exception is the brain. For example, it is well known that primates have a relatively larger brain than rodents.

In fact brain weight can vary as much as fifteen-fold for animals of the same size. The graph obtained by Sacher when he plotted log brain weight against log life-span. The scatter of the points around the regression line is rather less than. As the figure suggests, brain weight was found to be a better predictor of life-span than body weight. However, the best prediction of life-span was obtained by taking both body weight and brain weight into consideration. The regression line shown in the figure is based on the equation

$$x = 0.636w + 0.198y + 0.471$$

where x = log life-span, y = log body weight, and w = log (brain weight/body weight$^{2}/^{3}$). The term w is known as the *index of cephalisation* and is a measurement of relative brain size. As the figure shows, when both body weight and index of cephalisation are taken into account, tehre is no longer a clustering of the points for rodents below the line and for primates above it.

The observed life-span for man is that appropriate for an animal of his weight with such a relatively large brain. Sacher has also examined the relation between brain weight, body weight and longevity within different orders of mammals. In general, the data confirmed the hypothesis that length of life is an increasing function of brain size when body size is held constant. One interesting exception to the general pattern was found with the order Chiroptera, the bats, which have lifespans of up to 20 years.

This is about three times as long as would be expected on the basis of their brain and body weights. In order to understand the significance of this finding it is necessary to consider the possible explanations of the correlation. Sacher suggested that the body-weight term is important because metabolic rate is negatively correlated with body weight.

Since metabolic errors are inevitable and increase the probability of dying, large animals will, because of their lower metabolic rates, have a lower incidence of metabolic errors and hence longer life-spans than small animals. According to Sacher, the brain size of an animal is important because animals with larger brains have superior homeostatic regulatory mechanisms.

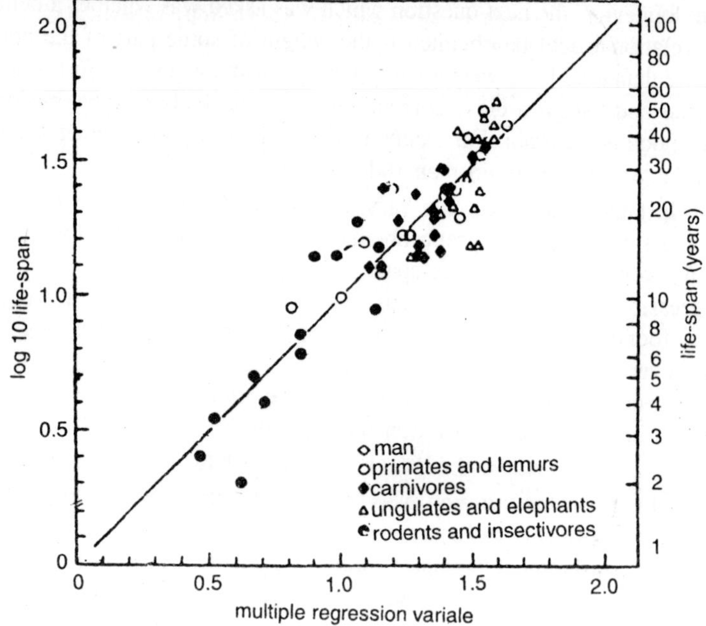

Figure 8.7: Relation of life-span to a combination of body weight and brain weight.

If a metabolic error causes the internal environment to fluctuate away from the optimum, the likelihood of this resulting in an irreversible change leading to death will be lower in animals with good regulatory mechanisms. Thus for animals of comparable size, the potential metabolic-error rates are similar, but the animal with the larger brain will live longer because its superior homeostatic control mechanisms minimise the effects of any fluctuations away from the normal levels.

If Sacher's interpretation of the body weight-brain weight-lifespan relation is correct, then the fact that bats do not show the same relation as other mammals is not surprising. Most bats have a period of diurnal torpor and also a period of hibernation which lasts for several months. During torpor and hibernation the metabolic rate is lower. Therefore, if longevity is inversely related to metabolic rate, they would be expected to live longer than mammals of comparable size and brain weight.

Sacher's hypothesis provides an explanation for differences in the life-spans of different species. Does it help to explain the longevity differences found within a species? When Storer (1967) looked at the brain weight, body weight, metabolic rate and life-span of 18 different strains of mice to see whether the relation that Sacher found between species also held within species, he found that it did not. Indeed he found that the metabolic rate was higher in the long-lived strains.

Although Sacher's hypothesis is attractive, it is not the only possible explanation of the observed relation between brain weight and longevity in mammals. An alternative interpretation of the correlation is that natural selection for long life and large brain size have gone together. For example, it has been argued that natural selection will favour increased brain size in potentially long-lived species, because large-brained animals have a greater learning capacity than small-brained animals.

An animal that can learn by experience is more likely to survive to old age than one which does not. Therefore selection pressures which favour increased longevity will also lead to increased brain size, because increased brain size enhances the chances of avoiding death.

A rather different explanation of the association between brain weight and longevity has recently been suggested by Sacher and Staffeldt (1974). They found that there is a strong correlation between gestation time and brain weight at birth. Mammals with large brains have long gestation times.

Sacher and Staffeldt believe that in order for large brains to develop, gestation times have to be long and litter sizes small. If this is so, selection for increased brain size would result in a decreased reproductive rate. A consequence of this might be that natural selection would favour an increased life-span, because this would allow a longer reproductive period to compensate for the decreased reproductive rate. Thus, according to this argument, the long life-spans of large-brained mammals could be the secondary consequences of selection for increased brain size.

The life-span of non-mammalian vertebrates

Some of the maximum-longevity records for non-mammalian vertebrates. The table is based on data given by Comfort (1964) who gives the source of each record. As with so many of the mammalian longevity records, the life-spans given are unlikely to be very reliable, because so few animals have been kept in captivity.

The data for birds come mainly from zoo records; as we discussed

Table 8.2 : Maximum recorded life-spans for non-mammalian vertebrates.

Species	*Common name*	*Maximum life-span (years)*
Birds		
Bubo bubo	eagle owl	68
Coracopsis vasa	vasa parrot	54
Pelicanus onocrotalus	white pelican	51
Aquila chrysaetos	golden eagle	46
Megalornis grus	grey crane	43
Larus argentatus	herring gull	41
Sarcorhamphus papa	king vulture	40
Anser anser domesticus	domestic goose	35
Columba livia domestica	domestic pigeon	30
Struthio camelus	ostrich	27
Goura cristata	crowned pigeon	16
Reptiles		
Testudo sumeiri	Marion's tortoise	152^{+}
Testudo marginata	margined tortoise	28
Terrapene carolina	Carolina box-tortoise	123^{+}
Clemmys guttata	Speckled terrapin	42^{+}
Caretta caretta loggerhead	turtle	33
Alligator sinensis	Chinese alligator	52
Eunectes murinus	anaconda snake	29
Anguis fragilis	slow-worm lizard	33
Sphenodon punctatus	tuatara	28^{+}
Amphibians		
Megalobatrachus Japonicus	giant salamander	52^{+}
Solamandra salamandra	European common salamander	24
Bufo bufo	Common toad	36
Xenopus laevis	African clawed toad	15
Rana temporaria	European common frog	12^{+}
Triturus pyrrhogaster	Japanese newt	25

in the previous chapter, it is very unlikely that many birds in the wild reach anything like their maximum potential longevity. In captivity, the life-spans of even the smaller birds seem to be long compared with those of mammals.

Comfort cites a case of a chaffinch living for 29 years in captivity, far longer than any mammal of a comparable size. Life-table data for birds in captivity are sparse, but in one study of the longevity of a group of 39 Bengalese finches the age at death of the oldest survivor was nearly ten years. It would be interesting to know whether the relation between the longevity, brain and body weights of birds is similar to that found by Sacher for mammals.

From the data available it appears that there is only a slight tendency for the larger birds to have longer lifespans than the smaller species. Reptiles are noted for their long life. There are several well-authenticated records of tortoises living for more than 100 years. According to Comfort, it is likely that many of the other reptiles are potentially long-lived although, because animals such as snakes are difficult to keep in zoos, the longevity records do not appear to be great. Data for amphibia are very limited but, in spite of the paucity of information, the maximum recorded life-spans of some of the smaller species are quite long.

There is rather more speculation about the longevity and senescence of fish than there is factual knowledge. Facts are of interest and importance, because it has been suggested that, unlike terrestrial vertebrates, fish show no senescence and never die a "natural" death. This conclusion was reached by Bidder than 50 years ago. He believed that many fish were capable of continued growth throughout life, and that their lack of senescence was associated with their indeterminate growth.

According to Bidder, since life on land sets a mechanical limit to size, the transition from an aquatic to a terrestrial environment involved the evolution in early land vertebrates of a size-regulator mechanism. This regulator mechanism produces a steady decrease in growth with age until the optimum adult size is reached, and Bidder suggested "that senescence is the result of the continued action of the regulator after .growth has stopped". Thus, according to this hypothesis, senescence is the by-product of the evolution of determinate size.

If Bidder's ideas are correct, then there are two types of vertebrate, one of which has a determinate size and shows senescence, and the other type which shows no senescence and has an indeterminate size.

Unfortunately information about both ageing and the growth rates of mature fish is not extensive. It is certainly true that many of the large teleosts continue to grow for very long periods of time, although the rate of growth tends to decrease with age.

The sturgeon continues grow for at least 30 years. It is also true that some fish do live for a very long time. Reliable reports suggest that ages of over 70 years are not uncommon for the sturgeon in natural conditions. Goldfish commonly live 17 years and have been reported to live for as long as 30-40 years in aquarium conditions. It is possible to obtain life-table data for natural populations of fish -since an individual's age can often be assessed by counting the annual growth rings in otiliths, scales, opercular bones or fin rays.

However, most of the populations which have been studied are for fish of economic importance and, since fishing affects survival chances in a non-random way, the data obtained are often difficult to interpret. The studies which have been made show that natural mortality in populations of large fish is very high, and survival curves which are obtained are often -of the exponential die-away type.

Although this type of curve is characteristic of populations of animals which do not show senescence, it is also found with animals such as wild birds, in which the death rate is so high that few animals live long enough for an increase in the force of mortality to become apparent. Life-table data from wild populations cannot, therefore, be taken as evidence that senescence does not occur in a particular species of fish.

Some fish life-tables do show a slight increase in the force of mortality with age. This is true for life-tables for the sturgeon, some herring and whitefish populations. A much more sudden and dramatic increase in the force of mortality is found for those species (such as the Pacific salmon) where nearly all individuals die after spawning for the first time.

Natural populations of fish therefore show a range of types of survival curve, from those which are very rectangular to those which are characteristic of animals with indeterminate life-spans. Thus there is in fact little evidence from life-table data to support the conclusion that absence of senescence is a characteristic of fish, although of course the possibility that some do not undergo an age-related deterioration cannot be ruled out.

Studies of physiological performance in fish of different ages and sizes would be valuable; at present, apart from evidence that some fish show a decline in reproductive capacity with age, little physiological information is available. The most detailed investigation of the relation between, growth and ageing of fish in aquarium conditions was made by Comfort (1964) who used the guppy *Lebistes reticulatus.*

In this species the maximum longevity is about 5 years; the mal e grows to a limited size, but the female shows indeterminate growth. Comfort was able to control growth by manipulating. the amount of food and space available to the fish. He found that, in all of the conditions which he used, he obtained life-tables which showed the pattern typical of senescence-the force of mortality increased with age.

Contrary to what would be expected from Bidder's hypothesis, senescence did take place while the guppies were still capable of growth. Since the growth characteristics of some species of fish living in natural conditions are known, Comfort was able to calculate the maximum longevity expected for some of them on the basis of the relation between growth and longevity that he found in the guppy.

Assuming that growth and survival in the stickleback, cod, hake and plaice are related in the same way as in the guppy, their growth characteristics lead to expected longevities of up to 5, 20, 40 and 60 years respectively. These values exceed those recorded for wild fish of these species. Therefore, if Comfort's extrapolation from the guppy to these fish is valid, it does suggest that evidence of senescence in natural populations should not be expected unless large amounts of reliable life-table data are available, because so few fish achieve anything like their maximum longevity potential.

We shall return to the subject of the relation between the rate of growth and ageing in the next chapter, since there is evidence from fish and other animals that longevity and senescence are in some way related to the rate of growth and development. However, although the evidence is far from conclusive, it seems unlikely that growth and ageing are related in the way that Bidder suggested, because there is little reason to believe that fish show no senescence, even if they do have indeterminate growth.

The life-span of invertebrates

Comfort (1964) has collected together longevity records for more than 250 different species of invertebrates from different phyla and classes. Table 8.3 gives some of these records; usually the longest life-span listed for each class or phylum has been chosen. The records for the different species are not really comparable because some are based on laboratory life-tables, some on observations of single individuals in captivity, some on direct observations of animals in their natural environment, and some on age estimates from growth rings.

Furthermore, it is well establishe that environmental factors such as temperature can have very marked effects on the longevity of invertebrates, so at best the longevity records gi,.en can provide only a very approximate idea of relative life-spans. It must also be borne in mind that the "life-span" often means different things for different species. For example, for many insects the life-span given is the survival time of the imago, yet frequently the length of the pre-imaginal stages is much greater than this, particularly for the species with non-feeding imagoes.

It is also difficult to decide on the life-span of animals such as some of the planarians which can divide by fission to form new individuals. It is impossible to survey in detail in this chapter the whole range of invertebrate life-cycles and life-spans. Comfort's book summarises and evaluates much of the available data, so our discussion here will be limited to a brief consideration of the important question

Is there any evidence that some of the invertebrates do not show senescence?

Comfort concluded that the strongest evidence for indeterminate life-spans came from data for the coelenterates. There are well-substantiated records of sea anemones living for more than 80-90 years. This very long life-span may be associated with the well-known ability of coelenterates to renew or replace worn-out or damaged parts.

Even if the extreme longevity of some actinians does reflect the absence of senescence in some species in this class, indeterminate life-spans are certainly not found in all coelenterates. For example, for some of the colonial hydroids, it is known that the life-spans of individual hydranths are limited.

The other phylum for which Comfort believes that "the indeterminacy of life-span cannot be excluded" is the Mollusca. Some of the smaller molluscs are known to have relatively short determinate life-spans : laboratory-kept populations of the pulmonate *Limnaea columella* show a typical rectangular type of survival curve with a maximum longevity of about 7 months.

Natural populations of the bivalves *Nucula turgida* and *Amphidesma ventricosum* have also been found to show an increased force of mortality with age, indicating that they too undergo senescent changes. In other species there is no evidence of senescence. Possibly the longest-lived invertebrate known is the fresh-water mussel *Margaritana margaritfera*

Table 8.3 : Maximum recorded longevities for some invertebrates

Phylum and Class	*Species*	*Maximum life-span*	*Evi-dence*
Porifera			
Demospongiae	*Suberites carnosus*	15 y	c
Coelenterata			
Anthozoa	*Cereus pedunculatus*	85-90 y	c
Platyhelminthes			
Cestoda	*Taeniarhynchus saginatus*	>35 y	h
Turbellaria	*Dugesia tigrina*	6-7 y	c
Aschelminthes			
Nematoda	*Wucheria bancrofti*	17 y	h
Rotifera	*Callidina sp.*	5 mo	c
Annelida			
Polychaeta	*Sabella pavonina*	>10 y	c
Oligochaeta	***Lumbricus terrestris***	5-6 y	c
Arthropoda			
Arachnida	*Filistata insidiatrix*	11 y	c
Crustacea			
S.C. Cirripedia	*Balanus balanoides*	>5 y	w
S.C. Malacostraca	*Astacus*	15-25 y	?
Insecta			
O. Ephemeroptera	*Cloeon dipterum*	4 w	c
O. Isoptera	*Neotermes castaneus*	>25 y	w
O. Lepidoptera	*Maniola jurtina*	44 d	c
O. Coleoptera	*Blaps gigas*	>10 y	c
O. Hymenoptera	*Lasius niger*	>19 y	c
Echinodermata			
Echinoidea	*Echinus esculentus*	> 8 y	w
Asteroidea	*Marthasterias glacialis*	>7 y	c
Molluses			
Amphineura	*Chiton tuberculatus*	12 y	w
Gastropoda			

Table Contd.

Phylum and Class	*Species*	*Maximum life-span*	*Evi-dence*
S.C. Prosobranchia	*Patella vulgata*	15 y	wg
S.C. Opisthobranchia	*Haminea hydotis*	4 y	w
S.C. Pulmonata	*Rumina decollata*	12 y	c
Bivalvia	*Margaritana margaritifera*	70-80 y	wg
Cephalopoda	*Loligo pealii*	3-4 y	w

Abbreviations

Age : y=years, mo=months, w=weeks, d=days.

Evidence : c=kept in captivity, w=in wild conditions, g=age estimate based on growth, h=host case history.

which, on the basis of shell growth rings, has been estimated to live for up to 100 years.

In theory, the shell structure of many molluscs enables estimates to be made of not only age, but also growth rates, so it might be thought that molluscs would be excellent material for investigating the relation between growth and ageing. However, the difficulty is that, in the absence of growth, the annual growth rings are not formed. Thus there is no adequate way of knowing if and when periods of arrested growth have occurred.

The studies of growth and ageing which have been made suggest that in many species growth continues up to the maximum recorded age. In fact, the position with regard to growth and senescence in molluscs seems to be very similar to that for fish: some species undoubtedly show senescence, but for others there is no reason to believe that it does occur, although the data are not really sufficient to justify the belief that they have indeterminate life-spans.

Similarly, although growth is known to be limited in some species, in others it does appear to continue throughout life. As with the fish, more information might provide valuable insights into the relation between growth and longevity.

The life-span of protozoan clones

The early literature about gerontology contains many papers dealing with the problem of whether or not protozoan cultures are immortal. It was generally believed that they were, but there were repeated reports of cultures which died out after a few months. It was realised that, as with so many other animals, failure to survive might indicate simply that inappropriate environmental conditions bad been provided, rather than that senescent changes had occurred.

Consequently, for many years the arguments continued with some people claiming that Protozoa were immortal, and others suggesting that they showed clonal senescence. The problem is still far from resolved, but for a few species experimental work has shown fairly clearly how in certain conditions some protozoan clones do undergo senescence, while in other conditions they do not.

Much of the experimental work has been concentrated on the ciliates, particularly *Paramecium aurelia*. It is now certain that *Paramecium* and several other ciliates show clonal ageing. In order to understand the nature of the ageing process shown by these ciliates, it is necessary to give a brief and rather crude outline of their methods of reproduction. *P. aurelia* will be used as an example.

The ciliates have two types of nuclei large polyploid macronuclei which control the general metabolism of the cells, and small diploid micronuclei which are important in sexual reproduction. The macronuclei can be thought of as controlling the somatic functions of the cell, the micronuclei as controlling the germinal functions. *P. aurelia* has one macronucleus and two micronuclei.

It can produce asexually by binary fission : each micronucleus divides mitctically to produce identical daughter nuclei, and the macronucleus also divides so that each daughter cell has two micronuclei and a macronucleus. In addition to simple fission, two types of sexual process are known for *P. aurelia :* conjugation which involves mating between individuals of complementary mating type, and autonomy which is a form of self-fertilisation.

During conjugation, the two micronuclei in each irdividual of a pair of mates divide meiotically so that 8 haploid nuclei are formed in each cell. Seven of these nuclei disintegrate, and the remaining one divides mitotically to produce 2 gamete nuclei in each cell. Reciprocal exchange of one of the gamete nuclei occurs between the two conjugating cells. The gamete nuclei then fuse, and the two conjugating cells separate. The new diploid nucleus undergoes two mitotic divisions to produce two new micronuclei and two new macronuclei; the original macronucleus has by this time disintegrated. At the next cell division each micronucleus divides mitotically, and one macronucleus passes to each daughter cell, thus restoring the original complement of two micronuclei and one macronucleus per cell.

The most important things to remember about this complex mating process are that it involves the exchange of genetic material between cells and the formation of a new macronucleus. Autogamy, the other type of sexual process, is very similar to conjugation in many ways, but no genetic material is exchanged between cells. The whole sexual process occurs within a single cell.

The types of nuclear divisions which occur are the same as those in conjugation, but instead of exchanging gamete nuclei with another cell, the two identical sister gamete nuclei within each cell fuse to form a new diploid nucleus. The processes by which this nucleus then divides to produce new micro- and macronuclei are exactly the same as those following the formation of the new diploid nucleus in conjugation. The old macronucleus in replaced by a new one.

Autogamy and conjugation do not normally occur in culture conditions which allow high asexual fission rates. If a clone has been initiated with a cell which has just been produced by autogamy or conjugation, it is usually capable of many generations of asexual reproduction- up to *350* fissions in some strains of *P. aurelia,* However, eventually, unless sexual reproduction is allowed, the fission rate declines, abnormal cells are produced, and the clone dies out. The clone does not have an indefinite life-span.

Conjugation or autogamy can be induced by semi-starvation, but clonal age affects the ability of individual cells to take part in sexual processes. During the first fissions of a newly initiated clone, the cells are incapable of conjugation or autogamy.

This early period of "sexual immaturity" is followed 'by a period of "maturity" in which cells will conjugate if they are given the appropriate conditions, but will not go through autogamy. Following maturity there is a transition period during which cells are capable of either conjugation or autogamy, but eventually "senescence" begins and the cells can undergo autogamy but not conjugation.

During senescence the viability of the progeny clones produced as a result of autogamy gradually gets less until in extreme old age it is very difficult to start a new clone. Thus the original clone is no longer able to found new clones by sexual processes, or to continue its own line of asexual fission. Senescent changes which are comparable, although not identical, with those found in *P. aurelia* have been found in other *Paramecium* species and also in other genera of ciliates, e.g., *Euplotes* and *Tetrahymena.*

The ciliates are not therefore immortal since, unless the nuclear reorganisation in%olved in conjugation or autogamy occurs, cultures invariably die out. The renewal of the macronucleus seems to be essential for the rejuvenation of a colony. In many ways, of course, the type of ageing and senescence shown by these ciliates is analogous to that shown by most Metazoa, where the somatic cells have only a limited life-span and the line can only be continued when sexual reproduction initiates a new individual with a new soma.

Although the clonal senescence shown by ciliates is interesting, it must be remembered that ciliates are unique in having, two distinct types of nucleus. Studies of Protozoa such as *Amoeba* are possibly more relevant to understanding ageing in Metazoa since, at least superficially, they resemble metazoan somatic cells.

The most commonly studied species, *A. proteus* and *A. discoides,* are believed to lack a sexual process; they divide by simple binary fission in a way which resembles the mitotic division seen in metazoan somatic cells. Do they show any evidence of senescence, or are all clones potentially immortal? Danielli and Muggleton (1959) have investigated this problem.

They found that, when they kept amoebae in culture conditions which provided a rich supply of the food source, the amoebae would grow and divide repeatedly. Under these conditions of logarithmic growth, the cultures could be maintained indefinitely. There was no evidence of death of individual cells, apart from that due to accidents. Therefore the cultures kept on this "growth" diet were immortal; they bad indeterminate life-spans.

If, however, amoebae were kept on a less-rich diet, a "maintenance" diet, they did not divide, and eventually died. The most interesting observation was the effect of keeping the amoebae on a maintenance diet for a few weeks, and then transferring them to the growth diet. They did not, as might be expected, resume growth and multiplication and continue indefinitely. Instead it was found that, although growth and division were resumed, the cultures had only a limited life-span of between 30 days and 30 weeks. In Danielli and Muggleton's terminology, they became "spanned". Their exact life-span depended on the conditions of the maintenance diet.

Two types of amoebae were present in the spanned cultures. One type was capable of normal logarithmic multiplication, with one cell dividing to give two daughter cells, each of which could divide in turn to give two daughter cells, and so on. Eventually, however, both of the daughter cells died. The other type of amoeba was also capable of repeated division, but each division resulted in one viable cell and one inviable cell until ultimately, after a number of divisions, both daughter cells died.

The reason why a short period on the maintenance diet converts an immortal culture into a culture with a limited life-span is not known, but these experimental results obtained with amoeba are worth remembering when thinking about the results obtained in *vitro* work with metazoan cells. What does seem to be clear from Danielli and Muggleton's experiments is that, even for sexual species the mortality of immortality of protozoan cultures depends in a rather subtle way on the culture conditions.

Ageing changes in renewing tissues

Tissues which are continually turning over might not be thought to be important in ageing, since defective or dead cells can be replaced. However, there is ample evidence to show that age-related changes do occur in the cells of these tissues. For example, it has been found for a number of tissues that the stem cell generation tine is different in old and young animals.

The mitotic cycle is conventionally divided into four phases. If animals are injected with tritiated thymidine, a radioactively labelled nucleoside which is incorporated into replicating DNA, cells which are in or are about to enter S-phase at the time of injection incorporate the labelled molecules. Unless thymidine is incorporated into DNA, it is broken down and lost from the body within a short time after injection.

Thus, by injecting animals with radioactive thymidine, it is possible to label a cohort of cells which were in S-phase at the time of injection. By killing animals at various times after injection, and examining their tissues using an autoradiographic technique to locate the sites of the labelled molecules, it is possible to follow the fate of this labelled cohort of cells.

Using this technique the cell generation time and movements of cells in a number of renewable tissues have been studied. Lesher and his colleagues have made extensive studies of the intestinal mucosa of mice. The stem cells for this tissue are located in the crypts of Lieberkuhn, where active mitotic division can be seen. Newly formed cells are pushed out from the crypts onto the villus, where they gradually pass to the tip and ultimate drop off.

The percentage of labelled mitotic figures found in the crypt cells of mice of different ages at various intervals after injection with tritiated thymidine. The stem cells do not form a synchronous population, so only a proportion of the cells were in S-phase during the time when tritiated thymidine was available for incorporation into the DNA. Animals which were killed $_{J}$ hour after injection showed no labelled mitotic figures, because the cells undergoing mitosis were in G_E at the time of injection. Gradually, however, 1-2 hours after injection, mitotic figures were found to be labelled because the labelled cohort of cells which were in S-phase began to undergo mitotic division. Between 2 and 79 hours, all of the cells which showed mitotic figures were labelled, i.e., all of the dividing cells were from the cohort of cells which was in S-phase when the animal was injected.

In the following 3-4 hours, the number of labelled mitotic figures fell because the cells which were in G_1 or M at the time of injection entered mitosis. There was then a gradual increase in the percentage of labelled mitoses as the cohort of labelled cells began its second division.

By looking at the time between corresponding points on the curves for the first and second waves of mitoses, it is possible to calculate the length of the complete mitotic cycle, i.e., the cell generation time. Estimates of the time spent in the different phases of the cycle can be calculated from the slopes of the ascending and descending limbs of the curves, cnd the levels and positions of the plateaus and troughs.

The figure shows that as mice get older there is an increase in the cell generation time. The biggest increase occurs as the juveniles become mature. There is little change between 30u and 675 days of age, but in old animals the generation time again increases. This increase in generation time is largely due to an increase in the amount of time that the cells spend in G, and S.

In addition to the studies of cell generation time, Lesher and his colleagues have shown that the transit time of cells from the crypts to the extrusion zone increases with age, and that the number of crypt cells which are dividing decreases from 132 cells per crypt in 100-day-old mice to 92 cells per crypt in 825-day-old mice.

They estimate that the decrease in the number of proliferating cells, combined with the increased generation time in old animals, means that the rate of production of new epithelial cells is 50% less than in a 3month-old mouse. Since there is no evidence of a decrease in the number of cells in the villi as the animal ages, this means that the life-span of newly formed epithelial cells must be greater in old animals.

What determines the longevity of the differentiated cells in he villi and the mitotic rate of the stem cells in the crypts? ire cell cycle-times and cell life-spans increased because the tem cells themselves are showing ageing changes, or do they ary as a result of changes in the extracellular environment e.g., hormone levels or the efficiency of the vascular supply)?

There is reason to believe that the stem cells are capable of faster mitotic rates than are normally found in old animals. In mice which are subjected to low levels of continuous irradiation, the length of the cell cycle is considerably reduced even in old animals. This is presumably a compensatory reaction to produce new cells to replace those which are damaged by radiation. The fact that in old animals the mitotic rate can be accelerated suggests that the increased generation time which occurs in normal ageing does not occur as a result of changes in the stem cells themselves.

The whole problem of what determines the mitotic rate of stem cells and the longevity of differentiated cells is an intriguing one. In an adult mammal, the rates of cell formation and cell loss must be delicately balanced. If they were not, there would be changes in the mass of organs and tissues. When the stem cell mitotic rate is high, as in the intestinal crypts, the life expectancy of the differentiated cells must be short. When the mitotic rate is moderate, as in the basal cells of the epidermis, the life-span of the differentiated cells must also be moderate. In tissues such as the liver, where mitotic rate is normally very low, the functional cells must be very long-lived.

Bullough (1967) has argued that the balance between cell loss and cell gain is maintained because both the mitotic rate and the rate of cell ageing depend on chalone control systems. Chalones are tissue-specific mitotic inhibitors which are synthesized within the cells of the tissues on which they act. Their action in inhibiting mitosis is enhanced by adrenalin and glucocorticoid hormones.

When chalone concentration falls, cells enter mitosis; when it is high, mitosis is inhibited and cells enter what Bullough calls the "ageing pathway" which leads to functional maturity and ultimately death. Bullough believes that chalone action determines not only the mitotic rate in tissues, but also the life-expectancy of the cells. He has pointed out that, when mice are stressed by partial starvation, the mitotic rate in the epidermis is greatly reduced, yet the thickness of the epidermis is not decreased.

This must mean that the rate of cell loss is also lower than in normal animals, i.e., the cells have a longer life-span. Conversely, it has been shown that in psoriasis where epidermal mitotic rate is greatly increased, the life-span of the cells is reduced from 14-21 days to only 4 days. Thus cell life-expectancy and mitotic rate are closely correlated-within any one tissue lifespan is inversely proportional to mitotic rate. The relation between mitotic rate and life-span for sebaceous gland cells in rats where the mitotic rate varied as a result of castration, hypophysectomy, and testosterone injection.

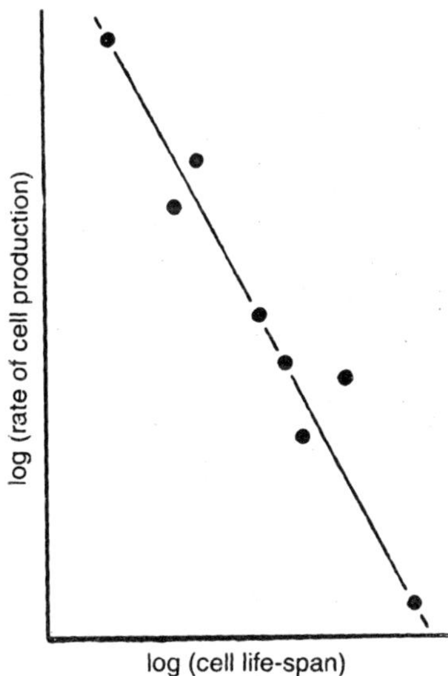

Figure 8.8 : Relation between the life-span and mitotic rate of rat sebaceous gland cells.

If it is accepted that chalone concentration controls both the rate of mitosis and cell life-span, does it do so because mitotic rate itself determines cell I ife- expectancy, or because chalones have two separate actions, one on the mitotic rate and one on the rate of ageing? Is cell life-span determined by the length of the mitotic cycle when it was formed? Bullough argues that it is not because, when cbalone concentration is changed, not only is mitotic rate in the stem cells changed, but the life-span of cells which have already differentiated is also altered. Therefore chalones affect the mitotic cycle and cell life-span independently.

Bullough (1973) speculated that in non renewable tissues, such as the brain and muscle, chalone concentration is very high and hence the cells age very slowly. Since it is known that in renewing tissues, stress hormones enhance chalone action, he suggested that stress may lead to an increased lifespan if non-dividing cells (such as nerve cells) are the pacemakers for ageing changes. Enhanced chalone action due to stress hormones could mean that fixed post-mitotic cells are more slowly. He suggested that the experiments in which the life-spans of mammals were increased by food restriction could be interpreted in these terms, since one of the effects of food restriction is to increase the size of the adrenals. Adrenalin enhances chalone action and therefore, according to Bullough's hypothesis it could decrease the rate of post-mitotic cell ageing, and thus increase the life-span of the mammals.

All of the renewable tissues which have been mentioned so far are ones in which the stem cells and fully differentiated cells remain in close proximity. There are other types of renewable "tissues", e.g., the blood cells in the circulatory system, in which the final products of differentiation do not remain closely associated with the stem cells. It could be argued that, in tissues such as the epidermis or gut epithelium, the life-span of the cells is determined by their position in the tissue. This cannot be true for cells such as red blood cells, which are derived from stem cells in the bone marrow.

Nevertheless, there is good evidence showing that red blood cells do show actuarial senescence, i.e., their destruction by macrophages is not a random process, but the likelihood of being destroyed increases as they get older. This has been shown by labelling a cohort of cells with radioactive chromium, injecting these cells into host animals, and following the loss of chromium from the blood. In man, the average erythrocyte life-span is about 120 days; in the rat it is about half this value.

During the life-span of red cells there are changes in enzyme activity, cell density, osmotic fragility an ' mechanical properties. Recently Kay (1975) has shown that senescent red cells seemed to be recognised and destroyed by macrophages because, although young cells are recognised as "self" by macrophages immunoglobulins in serum attach to the surface of old cells, thereby preventing the macrophages from recognising them as "self". Since they are no longer recognised as "self", the old cells are phagocytised.

Erythrocytes undoubtedly show senescence but, unlike the duodenal crypt cells discussed earlier, there is no reason to think that their life-span depends on the age of the animal. This has been shown by taking radioactively-labelled cells from young and old donor animals and following their fate in young and old hosts. The results showed that the cell lifespans are not affected by the age of the animal which produced them, or by the age of the animal which provides their environment.

However, although the life-spans of red cells seem to be unaffected by the animal's age, there is evidence that the quality of the cells may be different in old animals. For example, Silini and Andreozzi (1974) have reported that there are small changes in haematocrit, and that haemoglobin concentration in mouse erythrocytes decreases slightly with age. The causes of these changes are not clear.

Studying the mitotic rate and size of the stem cell population for erythrocytes is much more difficult than it is for epidermal or gut mucosal cells, and most attempts to determine whether or not blood-cell-forming tissue undergoes changes with age have been indirect. For example, Grant and LeGrande (1964) attempted to assess erythropoietic changes with age by measuring the level of circulating erythrocytes in rats which were kept at reduced atmospheric pressure or which had been made anaemic by removing a known amount of blood. Both of these treatments stimulate the bone marrow to produce new red cells.

They found that old rats took about twice as long as young rats to restore the red-cell level to normal after bleeding, and that in low oxygen tensions the number of circulating red cells did not increase to the same extent in old animals as it did in young. This may mean that there is an age-related decrease in the ability of the bone marrow stem cells to produce new erythrocytes, but the change could also be the result of a deterioration in the efficiency of the feed-back control system which regulates red-cell production. For example, the production of erythropoietin, the hormone formed in the kidney which is responsible for stimulating stem cells to differentiate, might be less efficient in old animals.

Another technique which has been used to assess the proliferative capacity of bone-marrow cells involves measuring their ability to form colonies in the spleens of irradiated mice. When mice are exposed to a high dose of radiation, their own proliferative cells are destroyed. If bone-marrow cells from another animal are suspended and a known concentration is injected into the irradiated host animal, some of the cells lodge in the spleen and divide to form small colonies which can he detected and counted.

The number of colony-forming units (CFUs) relative to the number of cells injected gives an estimate of the proliferative capacity of marrow cells. The results obtained using this technique have been contradictory. For example, Coggle and Proukakis (1970) found no age-related change in the proportion of CFUs for mouse-femur marrow cells. They also estimated that the total number of nucleated bone-marrow cells did not change with age, and hence that there was no change in the overall proliferative capacity of the blood-forming tissue.

Silini and Andreozzi (1974), on the other hand, found that femur marrow cells showed a decreased ability to form spleen colonies as age increased, but that there was an increase in the number of marrow cells in the femur.

Although the results were different, both of these studies suggest that there is probably no substantial change in the proliferative potential of the bone marrow with increasing age. It must be remembered, however, that the bone-marrow cells are pluripotential stem cells which can give rise to cells cornmitted to granulocyte, thrombocvte and lymphocyte production as well as to erythrocytes. Even if the bone marrow cells are still capable of proliferation, it does not mean that the capacity of old and young animals to produce erythrocytes is necessarily the same.

INDEX

Y

Z